MOBILE COMPUTING

*Implementing Pervasive Information
and Communications Technologies*

OPERATIONS RESEARCH/COMPUTER SCIENCE INTERFACES SERIES

Series Editors

Professor Ramesh Sharda
Oklahoma State University

Prof. Dr. Stefan Voß
Technische Universität Braunschweig

Other published titles in the series:

Greenberg, Harvey J. / *A Computer-Assisted Analysis System for Mathematical Programming Models and Solutions: A User's Guide for ANALYZE*

Greenberg, Harvey J. / *Modeling by Object-Driven Linear Elemental Relations: A Users Guide for MODLER*

Brown, Donald/Scherer, William T. / *Intelligent Scheduling Systems*

Nash, Stephen G./Sofer, Ariela / *The Impact of Emerging Technologies on Computer Science & Operations Research*

Barth, Peter / *Logic-Based 0-1 Constraint Programming*

Jones, Christopher V. / *Visualization and Optimization*

Barr, Richard S./ Helgason, Richard V./ Kennington, Jeffery L. / *Interfaces in Computer Science & Operations Research: Advances in Metaheuristics, Optimization, and Stochastic Modeling Technologies*

Ellacott, Stephen W./ Mason, John C./ Anderson, Iain J. / *Mathematics of Neural Networks: Models, Algorithms & Applications*

Woodruff, David L. / *Advances in Computational & Stochastic Optimization, Logic Programming, and Heuristic Search*

Klein, Robert / *Scheduling of Resource-Constrained Projects*

Bierwirth, Christian / *Adaptive Search and the Management of Logistics Systems*

Laguna, Manuel / González-Velarde, José Luis / *Computing Tools for Modeling, Optimization and Simulation*

Stilman, Boris / *Linguistic Geometry: From Search to Construction*

Sakawa, Masatoshi / *Genetic Algorithms and Fuzzy Multiobjective Optimization*

Ribeiro, Celso C./ Hansen, Pierre / *Essays and Surveys in Metaheuristics*

Holsapple, Clyde/ Jacob, Varghese / Rao, H. R. / *BUSINESS MODELLING: Multidisciplinary Approaches — Economics, Operational and Information Systems Perspectives*

Sleezer, Catherine M./ Wentling, Tim L./ Cude, Roger L. / *HUMAN RESOURCE DEVELOPMENT AND INFORMATION TECHNOLOGY: Making Global Connections*

Voß, Stefan, Woodruff, David / *Optimization Software Class Libraries*

MOBILE COMPUTING

Implementing Pervasive Information and Communications Technologies

SHAMBHU UPADHYAYA

SUNY at Buffalo

Buffalo, New York

ABHIJIT CHAUDHURY

Bryant College

Rhode Island

KEVIN KWIAT

Air Force Research Laboratory

Rome, New York

MARK WEISER

Oklahoma State University

Stillwater, Oklahoma

KLUWER ACADEMIC PUBLISHERS
Boston/London/Dordrecht

Distributors for North, Central and South America:
Kluwer Academic Publishers
101 Philip Drive
Assinippi Park
Norwell, Massachusetts 02061 USA
Telephone (781) 871-6600
Fax (781) 871-9045
E-Mail: kluwer@wkap.com

Distributors for all other countries:
Kluwer Academic Publishers Group
Post Office Box 322
3300 AH Dordrecht, THE NETHERLANDS
Telephone 31 786 576 000
Fax 31 786 546 474
E-mail: services@wkap.nl

 Electronic Services <http://www.wkap.nl>

Library of Congress Cataloging-in-Publication Data

Mobile computing : implementing pervasive information and communications technologies
  / edited by Shambu Upadhyaya ... [et al.].
    p. cm. -- (Operations research/ computer science interfaces series; 19)
  Includes bibliographical references and index.
  ISBN 1-4020-7137-X
  1. Mobile computing. I. Upadhyaya, Shambhu. II. Series.

QA76.59 .M645 2002
004.165--dc21   FEB 0 4 2004   2002026737

Contents

Part IV Computing

Contributors

Vishal Anand, SUNY at Buffalo
Bharat Bhargava, Purdue University
Sourav Bhowmick, Nanyang Technological University, Singapore
Prabuddha Biswas, Oracle Corporation
Ilyoung Chung, Purdue University
Supratim Dev, University of Illinois at Urbana-Champaign
Linda E. Doyle, Trinity College, Dublin
Song Han, Oracle Corporation
Michael Horhammer, Oracle Corporation
Joyce Lucca, Oklahoma State University
Sanjay Madria, University of Missouri, Rolla
Mukesh Mohania, Western Michigan University
Pradeep Kumar Mukherjee, Duncan Infotech
Charles Murray, Oracle Corporation
Donal O'Mahony, Trinity College, Dublin
Manika Kapoor, IBM India Research Lab
Chunming Qiao, SUNY at Buffalo
Abhinanda Sarkar, IBM India Research Lab
Ramesh Sharda, Oklahoma State University
Ting-Chung Tien, SUNY at Buffalo
Shambhu Upadhyaya, SUNY at Buffalo
Mark Weiser, Oklahoma State University

Preface

Virtual enterprises and mobile computing are emerging as innovative responses to the challenges of doing business in an increasingly mobile and global marketplace. These undertakings can be leveraged to capture new markets, customize the delivery of products and services, streamline and expand operations, and form business collaborations for the purpose of sharing resources and risks.

In a rapidly changing environment, it is critical to focus on the fundamental technological aspects that enable the concept of pervasive computing. This book is designed to address some of the business and technical challenges of pervasive computing that encompass current and emerging technology standards, infrastructures and architectures, and innovative and high impact applications of mobile technologies in virtual enterprises. The various articles examine a host of issues including the challenges and current solutions in mobile connectivity and coordination, management infrastructures, innovative architectures for fourth generation wireless and Ad-hoc networks, error-free frequency assignments for wireless communication, cost-effective wavelength assignments in optical communication networks, data and transaction modeling in a mobile environment and bandwidth issues and data routing in mobile Ad-hoc networks.

The book is essentially based on the technical theme of the first IEEE sponsored Academia-Industry Working Conference on Research Challenges, 2000 (AIWoRC'00). The conference held in Buffalo, New York in April 2000 carried the theme "Next Generation Enterprises: Virtual Organizations and Mobile/Pervasive Technologies". Three papers have been selected from the AIWoRC'00 Conference, which cover some of the business and technical topics pertinent to the above theme. The book has been enhanced by adding several articles on the architectural and technical aspects of mobile computing and communication.

About This Book

The primary audience for this book is industry practitioners, university faculty, independent researchers and graduate students. The articles have a mix of current and successful efforts, innovative ideas on providing the infrastructure support, and open problems—both conceptual and experimental. While some of the chapters in the book are significantly revised versions of the papers that appeared in the AIWoRC conference proceedings, the rest are new proposals from experts in the field. People in the academia as well as industry can benefit from this book. All the articles have gone through a peer review process. It is anticipated that the book will act as a single, consolidated source of information on the cutting edge of pervasive computing technologies.

The reader will find four different flavors of mobile and pervasive computing and technologies, namely, business and management, architecture, communication, and computing. The first three chapters focus on the business aspects of mobile computing and virtual organization. The fourth chapter lays out an architecture for a fourth generation wireless network. Chapters 5 and 6 are geared towards communication technology, both wireless and wireline. Chapter 7 is a taxonomy of data management environments in mobile computing and Chapter 8 is a review article on data and transaction management and research directions in this area. Finally Chapter 9 addresses various routing strategies for the seamless switching between mobile hosts in an Ad-hoc network.

Business and Management

The first chapter by P. Mukherjee opens with a review of the various digital cellular standards such as Global System for Mobile communications (GSM), Time Division Multiple Access (TMDA) and Code Division Multiple Access (CDMA), and the evolution of 1st, 2nd and 3rd generation of mobile technology and the technology options available in a developing country like India. The issues involved in introducing a new technology into a developing country is discussed along with critical success factors for choosing an application for use with mobile computing. Some of the initiatives taken in India to use the technology in an innovative fashion are explained and supplemented with a few case studies highlighting successful use of the technology. It observes that the Indian web and Internet companies simply cannot risk not going mobile.

The second chapter by J. Lucca, R. Sharda, and M. Weiser is on coordination of technologies in virtual organizations. For successful virtual organizations, the first and foremost issue is the identification of system level elements such as organizational memory, hardware, and software. Also a number of knowledge management tools such as database access, expert systems, and groupware must be in place. Network technology, which is responsible for the physical connection of the various modules is a critical component of virtual organizations. This chapter addresses the need for a cross-disciplinary approach to bring all the pieces together. The authors give a comprehensive review of current state-of-knowledge in these related areas. Also, a framework for linking relevant technologies and a closer inspection of the database segment are presented. The chapter concludes with a description of such a framework in place in a virtual organization, and research issues in this area.

The chapter by P. Biswas, M. Horhammer, S. Han, and C. Murray explores location-based services, mobile applications, and region modeling. The authors describe a framework for providing location-based services. That is, users' proximity to a plethora of information sources is coupled with their preferences. What is ultimately delivered is information that is individually tailored to not just *who* the user is but *where* that user is. The author's framework is based on a separation of information content and the providers of that content. Their framework allows the aggregation and selection of content from multiple providers for selectively targeting a customer's criteria. Bridging the gap between a service provider's application and the providers of other content such as maps, driving directions, and Yellow Pages, are service proxies. A management infrastructure is also put into place that gives the service providers control over which content provider is accessed. Organization and access of services can be according to geographic regions. These concepts are illustrated through demonstration applications where services are constructed using content from different providers. This chapter is also a rich source of outstanding research-worthy issues relating to location-based services.

Architecture

D. O'Mahony and L. Doyle describe in their chapter a future 4th generation wireless network that employs a general-purpose mobile node capable of running a large range of network applications. The node consists of a dynamic protocol stack that can be configured, from a suite of protocols, based on the requirements of the underlying network and can adapt its mode of operation to the radio environment and prevailing wireless network architecture. The authors have created a software and hardware architecture that allows them to experiment with many different facets of wireless networking such as multiple wireless links, medium access control protocols and Ad-hoc routing protocols. The architecture allows exploration of security issues and multimedia applications for the system. The resulting node architecture is one that embraces flexibility and adaptability at all levels from the hardware front-end through software radio to the application layer and is also an architecture that facilitates multiple levels of security.

Communication

The next two chapters address the communication issues. S. Deb, M. Kapoor, and A. Sarkar look at the interference problem in wireless communication systems. They propose a scheme in which the base station tries to predict whether transmission to a particular user device on a particular frequency will encounter errors. Their algorithm extends previous work in this area, with its application to wireless systems, which employs a fixed frequency hopping sequence. Simulation results show that the proposed scheme results in substantial improvement in both throughput and goodput, particularly in error-prone environments. V. Anand and C. Qiao describe heuristic algorithms for finding optimal routing and wavelength assignments in optical networks that maximize profits. Fiber-based optical networks are the popular choice for backbone of wide area networks. These networks employ wavelength routing to establish all-optical data-path. The optimization problem is formulated as an integer programming problem. Different algorithms are described that solve both the off-line and on-line version of the routing problem. It is shown that the profit optimization objective as used here results in solutions different from the traditional approach of minimizing cost or maximizing throughput.

Computing

I. Chung, B. Bhargava and S. Madria give a taxonomy of data management techniques in mobile computing. In conjunction with the existing computing infrastructure, data management for mobile computing has given rise to significant challenges and performance opportunities. Most mobile technologies physically support broadcast to all mobile users inside a cell. In mobile client-server models, a server can take advantage of this characteristic to broadcast information to all mobile clients in its cell. The authors discuss topics such as data dissemination techniques, transaction models and caching strategies that utilize broadcasting medium for data management. There is a wide range of options for the design of model and algorithms for mobile client-server database systems. The proposed taxonomies provide insights into the tradeoffs inherent in each field of data management in mobile computing environments. S. Madria, B. Bhargava, M. Mohania and S. Bhowmick look at the implementation of database systems in wireless systems in their chapter "Data and Transaction Management in a Mobile Environment", survey fundamental research challenges particular to mobile database computing, review some of the proposed solutions, and identify selected upcoming research challenges. Potential research directions are also explored, including mobile digital library, data warehousing, workflow, and e-commerce.

The final chapter by T. Tien and S. Upadhyaya deals with connectivity, routing, wireless/terrestrial interfaces, real-time content transmission, and Ad-hoc networks. Connection-oriented communication requires maintenance of a route for a certain period of time. Under these circumstances, route switching is oftentimes performed among communicating units so that a connection between the source and destination units is maintained. This can be a vexing problem in highly mobile networks that will exhibit a high rate of unit migration. These authors propose monitoring the signal strength of messages, so that a unit in a route that receives an incoming message can detect possible route fluctuations locally and initiate discovery of a substitute route. This briefly synopsizes the breadth and depth of the problem addressed. In relation to the theme of this book, it should be noted that mobile products are increasingly called to support distributed organizations. Given the accompanying push for device portability, we become confronted with battery technology as being a key factor; yet batteries are not yet good enough. Their capacity is a constraining factor on many mobile devices and the rate of improvement in battery development continues to lag significantly behind other technologies. Thus the connectivity solution presented in this chapter is especially relevant because it is sensitive to signal strength problems that can be induced, not only by the distance between units, but by a unit's battery-power.

A Final Word

Finally, in taking a perspective of history as if being viewed from the future, readers may someday look upon this book as a microcosm of a stage in technological development that brought them to their present state. At that time, this text may be regarded as having delineated the baseline for virtual enterprises and mobile computing at the onset of the 21st Century, and for having posed important challenges that were subsequently met with an indefatigable urge for progress.

Shambhu Upadhyaya, SUNY at Buffalo, Buffalo, New York
Abhijit Chaudhury, Bryant College, Rhode Island
Kevin Kwiat, Air Force Research Laboratory, Rome, New York
Mark Weiser, Oklahoma State University, Stillwater, Oklahoma
Guest Editors

Acknowledgments

The concept for this book originated in early 2000 during the organizational stages of the IEEE Conference: "Academia-Industry Working Conference on Research Challenges (AIWoRC'00)" in Buffalo, New York. The organizers Prof. H. Raghav Rao and Prof. Ramaswami Ramesh felt that edited books on the theme of the conference would provide the authors of those papers dealing with targeted technologies the opportunity to write on their research at greater length and depth than allowed by the page limit necessarily imposed by the conference. In addition to disseminating more on the outcomes specifically from the conference, submissions for the book were to be solicited from others who, although they may not have participated in the conference, are known to be addressing the same research challenges. The idea sounded right and they arranged a meeting with Prof. Ramesh Sharda, Series Editor, Kluwer Series in Operations Research/Computer Science Interfaces, who was attending the conference. The guest editorial committee was put together and a proposal was written and submitted to Kluwer Academic Publishers in July 2000.

The guest editorial board likes to thank Raghav Rao and Ramaswami Ramesh for seeding the idea for this book. Ramesh Sharda's help in forming the guest editorial committee and his support in writing the proposal are gratefully acknowledged. This book would not have been possible without the coordination of Gary Folven, the Publisher and his support staff at Kluwer. The authors have done a tremendous job of revising their prior submissions or writing new contributions to this book. Their prompt responses throughout this project are appreciated. We got the help of many peer reviewers to get quality feedback on the manuscripts. We would like to thank them for their efforts. The help of Ting-Chung Tien and Hung-shu Wu in typesetting the manuscript is gratefully acknowledged.

Chapter 1

MOBILE COMPUTING – THE INDIAN VISTA

Pradeep Kumar Mukherjee
Duncan Infotech

Abstract: This chapter explains the status of mobile telephony from the Indian perspective. It explains different digital cellular standards e.g., GSM (Global System for Mobile Communications), TDMA (Time Division, Multiple Access), CDMA (Code Division, Multiple Access), the evolution of the 1st, 2nd and 3rd generation of mobile technology, and the technology options available to India. The issues involved in introducing a new technology into a developing country like India are discussed, along with critical success factors for choosing an application for use with mobile computing. Some of the initiatives taken in India to use the technology in an innovative fashion are explained and supplemented with a few case studies highlighting successful use of the technology.

Keywords: Access Protocol, Applications, Mobile Telephony, Solutions, Technology, Wireless Standards

1. THE NEW HORIZON

The path-breaking technology of the '90s, the Internet, has heralded a dramatic transformation in the business landscape and touched the life pattern of common people. Closely following it, the wireless Internet, or more specifically, mobile computing, is having a profound impact on our lives now. Unlike most technologies whose impact was first felt by the developed world, Internet and mobile computing are benefiting the developing countries by jump-starting them into the digital economy. In spite of the infrastructure and bandwidth limitations, developing nations, like India, are making the most of their early starter's advantage to catch up with the developed countries. Mobile Computing has opened up a new horizon for business and government, and India with its rich pool of IT expertise is poised to assume a leadership position in this arena.

With the rapid convergence of computing and communication industries, cell phones and Personal Digital Assistants (PDAs) are quickly evolving into powerful multi-purpose devices. These not only communicate and organize contact addresses and to-do lists, and check, send and receive e-mails and faxes, but also provide driving directions, Internet surfing capability, which enable users to participate in live auctions, buy and sell stocks, and check flight schedules. Soon enough parameters of agricultural soil will be monitored from a remote location and a doctor will be able to diagnose critical medical data from his residence and suggest a course of treatment. Limiting wireless technology to communication alone is myopic. Mobile Computing is in a position to play a significant role in the field of agriculture, business and healthcare, even more so for developing countries like India.

1.1   The Initiation

India was initiated into mobile telephony in 1996, and since then there has been a significant acceptance and use of the technology by a large cross section of the market. While the Internet usage has grown to 1.1 million users (according to Internet Service Providers Association of India, ISPAI), the Cellular Operators Association of India (COAI) has estimated 2 million mobile subscribers, a number that is currently growing at a rate of 100 thousand users per month. This rate of growth in the mobile devices market is almost as high as the rate of PCs being sold in the country. New communication technologies like GPRS (General Packet Radio Services) and UMTS (Universal Mobile Telephone Services) will further boost the efficiency of mobile devices by enabling faster and clearer communication of data, pictures and video. In addition to this, the SIM (subscriber identification module) card is likely to become the smart "mobile money" that powers m-commerce.

The penetration of mobile computing has been limited compared to the opportunity that India offers. The majority of mobile subscribers are from the major cities and the high-income group. However, this trend is changing as mobile telephony is making inroads into the lower strata of society and the smaller cities and towns. The growing popularity of the Internet in rural areas is opening access to farmers so that they may trade their cattle, look for cold storage facilities for potatoes, and auction their produce. It will not be long before mobile access to Internet will be the most sought after technology.

1.2   Growing Activity

There is a significant level of activity in India regarding Mobile Computing solutions. Multinational corporations like Ericsson, Nokia and Motorola have already established a strong base in India in this evolving technology. Some of the mobile cellular operators have also introduced their messaging products in the market. Meanwhile, many Indian start-up companies, like Integra, Micro, and SAS, are competing actively with their value added services in the market. System Integrators, like e-Capital, Planetasia.com, IBM, and CMS are taking an active part in bringing forward new applications in the market.

WAP (wireless application protocol) enabled sites carrying local contents in India are mushrooming. The number of companies working on different facets of this emerging technology is growing rapidly. Tata Cellular, one of the mobile services operators in Andhra Pradesh, a progressive state in India, has already introduced mobile Internet services. Service providers like www.sharekhan.com and www.waphoria.com are providing access through an ISP's (Internet Service Provider) network independent of the mobile operator's gateway.

1.3   Primary Issues

While there is a lot of activity in this segment, the critical issues related to technology, infrastructure and profitability still need to be addressed in order to make Mobile Computing accessible to the mass market.

The Instruments: The subscribers have to dial-in to a WAP site with a WAP phone. Unlike Internet browsers, WAP browsers cannot be downloaded and installed into every handset. Mobile phone vendors have not yet revealed a strategy to upgrade the older phones to WAP capability. Some of the phones with in-built modem can probably accommodate the micro browser, but the idea still needs to be tested. In addition to this, the older models have screen limitations. This means that almost the entire population of Indian mobile subscribers is inaccessible to the WAP sites, even if the mobile operator, or an independent web site operates a WAP service. The subscribers to these services will be old subscribers who have made a change over to WAP phones or new subscribers who fortunately found a WAP phone when they decided to go mobile. Even if the new breed of mobile subscribers decides to dabble with mobile Internet, the group is likely to be limited in number since WAP phones are expensive, priced between $400 (Rs.18,000) to $700 (Rs.30,000). Such pricing limits the number of subscribers.

Low Bandwidth: The next disturbing factor for mobile operators is the issue of where WAP would fit in the new scenario of General Packet Radio Services (GPRS), Edge, and 3G (third generation) cellular services. Some of the major Indian mobile operators are considering implementing GPRS in their networks. Cellular providers like AirTel, Delhi and BPL Mobile, Mumbai, have announced their intention of supporting GPRS. Since WAP is an independent protocol over advanced access technology such as GPRS, Edge or 3G, investments made on WAP sites and gateways remain protected by introduction of new access channels. However, it is not clear what will happen to the WAP phones that subscribers have bought at this juncture to access Mobile Internet. Will their present WAP phone work when GPRS is introduced? Most likely they will have to buy a GPRS phone in order to connect to the WAP sites much faster. So the question is, would subscribers pay a premium for a WAP phone when they can wait for a phone that can access the WAP sites at a much faster speed? No mobile handset vendor has answers at this time.

Tariff: The third roadblock is the issue of tariff. How does the WAP service operator charge the subscribers? Will it be free like any Internet web site? Or would some of the sites charge money for providing value-added services. If yes, how will the cost for the content be derived? On the other hand, if the mobile operators continue to charge money for services, will they still remain attractive enough to subscribers? Are these reasons enough to dissuade mobile operators and Internet companies from trying out WAP services? The cellular infrastructure providers are aware of the time lag between suitable handsets being available in the market and the network being up and ready with the features introduced. Regardless of whether enough phones are available, all mobile operators and Internet service/content providers are scrambling to get the WAP services up and running. The players are bullish and the stakes run in billions of dollars. Huge businesses are expected to develop from the need to run mobile portals, sell solutions to players setting up these portals, and sell mobile phones to access these mobile portals.

1.4   The Government's Role

Although the wireline phone system in India has been available for more than a century, little technological advancements were achieved until communications were deregulated about five years ago and Indians were introduced to new communication technologies such as cellular telephony, paging, e-mail, and Internet. In terms of telephone network and usage, India has little to boast of, but because of this, it does not need to be concerned about legacy networks when it comes to introducing new communication technologies. As the wireline networks get revamped, the government needs to push the wireless and multimedia industries by promoting technologies such as WAP. It is welcoming to see that the government today has changed its conservative outlook, understood the urgent need of a much more reliable communications infrastructure and embraced technology at a better pace. According to COAI, the mobile operators trade group, Mumbai, the business capital of India, ranks highest in number of users, with more than 400,000 users out of the 2.88 million users in India. The government made a smart decision to choose GSM as the technology for mobile cellular services. Aside from a few remote areas, the mobile operators have put up GSM networks in almost every part of the country.

2. TECHNOLOGY AND STANDARDS

2.1 Digital Cellular Standards

Three main digital cellular standards [5] indicated below are competing in the world market (see Table 1 for a summary).
1. Europe's Global System for Mobile communications (GSM)
2. U.S'. IS-136 standard and its IS-54 predecessor (referred as TDMA - time division, multiple access)
3. Qualcomm's IS-95 standard (referred as CDMA for code division, multiple access)

In addition there is Personal Digital Cellular (PDC), a regional standard used in Japan. According to Gartner report [6], the 15 million analog subscribers worldwide in 1998 will decrease to 3 million in 2003. While W-CDMA based networks has been the prominent standard in South East Asia, India has adapted GSM, which is, and will continue to be, the dominant standard worldwide. CDMA and TDMA handsets will represent the next largest segments. In Asia/Pacific, the world's most attractive market for mobile computing, China, is leading with 104 million subscribers while South Korea is second, with 23 million. Closely following the two is India, with 15 million subscribers. India is expected to become the fourth-largest market for GSM by 2001, with a large base of analog subscribers migrating to digital systems.

2.1.1 1G-2G-3G: The Generations

India is in the second generation of mobile communications [7] and has adopted GSM out of a debate in 1990. In the mid '90s, US went for PCS (Personal Communication Services), while the European Union and many other countries adopted GSM, which now has over 300 million users and allows worldwide mobility (Ref Fig. 1).

Table 1. Digital world standards

The Move to Higher Speeds

Technology	Current Data Rate	Emerging Data Rate	Future Evolution
CDMA			CDMA 2000
-Packet	14.4	144 Kbps	
-Circuit	14.4	115.2Kbps	
TDMA			EDGE
-Packet	NA	43.2 Kbps	
-Circuit	9.6	115.2 Kbps	
GSM			EDGE
-Packet	NA	170 Kbps	
-Circuit	14.4 Kbps	115.2 Kbps	

1G to 2G to 3G Evolution

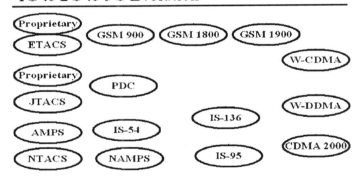

Figure 1. Generations of mobile communications

2.1.2   3G – The Ultimate Goal

3G is going to be the catalyst for a whole new set of mobile services, from mobile multimedia to machine-to-machine communications, thereby enabling access to advanced services anywhere, anytime. 3G brings together high-speed radio access and IP-based services into one, powerful environment, free from the confines of cables, fixed access points, and low connection speeds. Making 3G a reality depends on technology developments in different areas, including amendments in the core network and in the radio interface to support wideband communications. Supporting technologies such as WAP and Bluetooth also plays an important role.

3G brings together two powerful forces: wideband radio communications and IP-based services. Together, these lay the groundwork for advanced mobile Internet services, including personalized portals, infotainment, mobile commerce, unified messaging-encompassing high-speed data, superior quality voice and video, and location based services. The move towards IP is vital. Packet-based IP enables users to be "online" at all times, without having to pay until data is actually sent or recieved. The connectionless nature of IP also makes access significantly faster. Basically, 3G is about wideband radio communications with access speeds of up to 2Mbit/s.

2.1.3   The Migration Path – Why 3G?

The path from today's mobile technology to 3G is clearly mapped out and, in many cases, enables operators to retain much of their existing investment.

In networks based on CDMA the most likely route to 3G services is CDMA2000. CDMA2000 not only boosts the data rates of CDMA networks and provides a broader range of services, but also doubles the voice capacity of today's CDMA networks. The International Telecommunications Union (ITU) is likely to approve CDMA2000 as a 3G standard.

In networks based on GSM, the most widely deployed digital mobile standard, the most likely route to 3G services is EDGE. In the case of CDMA, the likely route would be either EDGE or WCDMA. The International Telecommunications Union (ITU) under IMT-2000 has approved of both EDGE and WCDMA. Both have their advantages, and it is expected that many operators will choose to adopt both solutions, providing an even broader range of possibilities and, most importantly, boosting overall network capacity.

2.1.4   GPRS and EDGE- India's route to 3G

Together, General Packet Radio Service (GPRS) and Enhanced Data for GSM Evolution (EDGE) mean that operators with 2G networks today can evolve to 3G services on existing network frequencies, using current network infrastructure. This means that operators do not need additional higher frequency bands such as those required for UMTS and can furthermore simply build on their existing investments, through a few straightforward hardware and software investments. However, both GPRS and EDGE would require new handsets to support.

For GSM operators, the first step is to implement GPRS, which introduces support for packet-based IP communications in the network. GPRS immediately increases user bandwidth and enables users to be "always on," which only paying for services when they are sending or receiving data. The next logical step would be to deploy EDGE, which enhances the radio element of the network to support 3G services at up to 384 kbit/s. Indian operators are most likely to adopt this route.

2.2   Frequency Availability

Region	Frequencies [8]
Japan	1918.8 – 2025 MHz and 2110 – 2200 MHz
South Korea	1885 – 2025 MHz and 2110 – 2200 MHz
Europe	1900 – 1996 MHz, 2010 – 2025 MHz & 2110 – 2186 MHz
United States	1850 – 1990 MHz for PCS; 2110 – 2200 MHz is reserved
ITU	1885 – 2025 MHz and 2110 – 2200 MHz

2.3   SIM Toolkit

In the initial realization of GSM, the SIM card played an essentially passive role, providing the user with the necessary authentication to access the network and storing the GSM encryption algorithms that ensure speech security. The oldest in mobile technologies, the SIM Toolkit extends the role of the SIM card, making it a key interface between the mobile terminal and the network. Using the SIM Toolkit, the SIM card can be programmed to carry out new functions. These include the ability to manipulate the menu structure of the mobile terminal to provide new, tailored options; for example, the handset could provide a menu for "domestic" use and a menu for business use. Either way, the phone becomes personalized to the individual and therefore user-friendly.

This technology has been used for wireless sports betting sites in Asia. Users are asked a series of questions regarding the betting and for each question a numerical answer needs to be entered from a list of options. Approximately three seconds later, the users confirm their choices and the bet is complete. The service is a clear improvement over WAP's snail-like sending and downloading speeds, and there is no need to log into the system. Data transfers happen in the form of SMS, a text-messaging capability that is built into the carrier's existing networks, but there are certain limitations with this technology. Since manufacturers are forced to place only one application or service on a given SIM card, users are not able to sample offerings from multiple companies without swapping cards.

2.4 Wireless Access Protocol

To understand Mobile Computing and Mobile Internet, it is necessary to appreciate the technological paradigms. Wireless Application Protocol (WAP) is a global, open standard that gives mobile users access to Internet services through handheld devices. It enables users to easily access a whole range of Mobile Internet and other data services from mobile devices such as smart phones and communicators, without the need to plug into a separate Laptop or data-enabled device. The built-in "WAP" micro browser allows information to be accessed directly from a phone in the same way that Web browsers provide access to on-line services via an Internet-ready PC. Typically, a WAP screen displays a number of hyperlinks to various services or information portals.

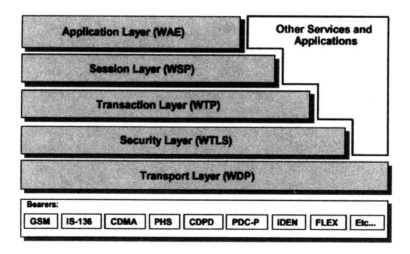

Figure 2. WAP protocol stack

2.4.1 WAP Architecture

The WAP architecture [9] shown in Fig. 2 provides a scaleable and extensible environment for application development for mobile communication devices. This is achieved through a layered design of the entire protocol stack. Each layer of the architecture is accessible by the layers above, as well as by other services and applications.

WAE (Wireless Access Equipment) includes a micro-browser environment with the following functionality:

- Wireless Markup Language (WML) - similar to HTML, but optimized for use in hand-held mobile terminals
- WMLScript – a lightweight scripting language, similar to JavaScript
- Wireless Telephony Application (WTA, WTAI) – telephony services and programming interfaces
- Content Formats – a set of well-defined data formats, including images, phone book records and calendar information.

The **Wireless Session Protocols (WSP)** consist of services suited for browsing applications (WSP/B). WSP/B provides the following functionality:

- HTTP/1.1 functionality and semantics in a compact over-the-air encoding
- Long-lived session state
- Session suspend and resume with session migration
- A common facility for reliable and unreliable data push
- Protocol feature negotiation

The protocols in the WSP family are optimized for low-bandwidth bearer networks with relatively long latency. WSP/B is designed to allow a WAP proxy to connect a WSP/B client to a standard HTTP server.

The **Wireless Transaction Protocol** (WTP) runs on top of a datagram service and provides a lightweight transaction-oriented protocol that is suitable for implementation in "thin" clients (mobile stations). WTP operates efficiently over secure or non-secure wireless datagram networks and provides the following features:

Three classes of transaction service:

- Unreliable one-way requests
- Reliable one-way requests
- Reliable two-way request-reply transactions
  Optional user-to-user reliability - WTP user triggers:
- The confirmation of each received message

Optional out-of-band data on acknowledgements

- PDU concatenation and delayed acknowledgement to reduce the number of messages sent
- Asynchronous transactions.

WTLS is a security protocol based upon the industry-standard Transport Layer Security (TLS) protocol, formerly known as Secure Sockets Layer -

SSL. WTLS is intended for use with the WAP transport protocols and has been optimized for use over narrow-band communication channels. WTLS provides the following features:

- Data integrity – WTLS contains facilities to ensure that data sent between the terminal and an application server is unchanged and uncorrupted.
- Privacy – WTLS contains facilities to ensure that data transmitted between the terminal and an application server is private and cannot be understood by any intermediate parties that may have intercepted the data stream.
- Authentication – WTLS contains facilities to establish the authenticity of the terminal and application server.
- Denial-of-service protection – WTLS contains facilities for detecting and rejecting data that is replayed or not successfully verified. WTLS makes many typical denial-of-service attacks harder to accomplish and protects the upper protocol layers.
- WTLS may also be used for secure communication between terminals, e.g., for authentication of electronic business card exchange.

Applications are able to selectively enable or disable WTLS features depending on their security requirements and the characteristics of the underlying network.

Wireless Datagram Protocol (WDP) the transport layer protocol operates above the data capable bearer services supported by the various network types. As a general transport service, WDP offers a consistent service to the upper layer protocols of WAP and can communicate transparently over one of the available bearer services. Since the WDP protocols provide a common interface to the upper layer protocols, the Security, Session and Application layers are able to function independently of the underlying wireless network. This is accomplished by adapting the transport layer to specific features of the underlying bearer. Global interoperability can be achieved by using mediating gateways to keep the transport layer interface and the basic features consistent.

2.4.2   Implementation of WAP Service

Implementation of a WAP-based solution requires three main components:

a. Direct-access server: Terminates the device's data call within the mobile network, which can then be routed over the Telco's Wide Area Network (WAN) or by leased line to an enterprise.

b. Gateway/proxy server: Converts WAP on the mobile side to Web protocols, which can then be routed over the Internet or Telco's Intranet. Thus, the Wireless Session Protocol (WSP) and Wireless Transaction Protocol (WTP) are mapped to HyperText Transfer Protocol (HTTP) and Transmission Control Protocol (TCP). Other features include end-user authentication, compilation of WML scripts, caching of data, and a Wireless Telephony Application (WTA) interface to telephony services like call control and handling. WML pages can be encoded and decoded from sources such as text or the WTA interface. Operation and Management (O&M) functions typically include service provisioning, fault management often via a Simple Network Management Protocol (SNMP) agent, performance management, configuration management, and billing support.

c. Application server - Connects to gateways and provides a platform for applications and Wireless Markup Language (WML) content, including run-time and development environments. If access is via the Internet then firewalls need to be installed for security between the gateway and application servers.

Enterprises wishing to WAP-enable their e-business applications, need to translate their HyperText Markup Language (HTML) web pages into WML. Standard XML authoring tools can be used to develop WML applications; in addition, the major vendors supply WAP Software Development Kits (SDKs), in some cases free of charge. The WAP specification uses HyperText Transfer Protocol (HTTP) version 1.1 to communicate between the WAP gateway and the Web servers, meaning that virtually any off-the-shelf Web server may be employed. Dynamic applications can also be devised using standard techniques, such as Active Server Pages (ASP), Java Server Pages (JSP), and Common Gateway Interface (CGI) scripts. Both WML and HTML pages can exist side-by-side on the same Web server.

Since most enterprises may be relying on Telco's gateway for mobile access, they need only to install a WAP application server in order to implement a wireless solution. However, enterprises such as banks that prefer users, for security reasons, not to traverse an external gateway, or that require a dial-in capability to their own Intranet, will also need to install a gateway. The major vendors of gateway and application servers, such as Ericsson, Nokia, Motorola, Phone.Com, and CMG, are targeting their products at the Telco, ISP, and enterprise markets. In addition, some vendors are concentrating on offering a combined end-to-end solution aimed at high-performance usage within enterprises.

Wireless Portals

These are specialized application servers aimed at Telcos where existing web content in a variety of formats can be transformed into native XML, which may then be rendered in the preferred mobile device format (eg. Oracle's Portal-to-Go, Lucent's Zingo and Motorola's MIX). Motorola's Voice extension Markup Language (VoxML) is an interesting format that works in conjunction with a voice browser, which allows Interactive Voice Response (IVR) to recognize a user's spoken request and uses text-to-speech synthesis to "play" the retrieved VoxML page.

2.4.3   WAP Service – The Functioning

The mobile WAP device is attached to the mobile network (GSM, CDDA, etc), which dials the dial-in server (RAS, or Remote Access Service). This server gives the WAP device access to the necessary protocols. These are the same lower level PPP protocol as are given in normal Internet service. These protocols are used to access the next step in the chain, the WAP gateway, hosted by the mobile operator. The WAP gateway connects the WAP device to the Internet.

The process is as follows: when a URL for site is entered on a WAP device, the WAP device first checks if it already has an open connection, and, if not, it dials-up the PPP provider or RAS. After the RAS has given the WAP device the required protocols and assigned it an IP address, the request for the URL is sent to a gateway. The WAP gateway, now under "control" of the WAP device, requests the URL with a normal HTTP request.

On the Internet, there is "web" server, which hosts both WAP and "web" contents, which receives the request to send out the contents located at the URL. The web server, depending on which type of browser it is talking to (WAP / "web"), sends out WAP, or "web" content as shown in Fig. 3, right hand side.

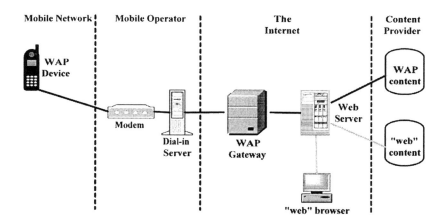

Figure 3. WAP – Functional setup – 1

Tracing the requested content back to the WAP device, the WAP gateway compiles the textual WML into so called tokenized WML, or WMLC (i.e "compressed" down into binary data). This tokenized WML is then passed back to the WAP device. If the contents from the web server are already in tokenized WML format, the WAP gateway skips this operation. The reason for the conversion from textual WML to tokenized WML is to reduce bandwidth usage. A WAP device's WML browser can only read tokenized WML.

Finally, at the WAP device that requested the URL, the WML browser, receives the tokenized WML code and renders the contents on the WAP device's display.

When the WAP device is configured to use a public WAP gateway, as opposed to one that is hosted by the mobile operator, the gateway is not behind a firewall as in Fig. 3. If the content provider requires control over the stream of data sent back and forth between the web server and the WAP device, a WAP server is installed. This device serves the dual role of a web server and a WAP gateway and is usually located behind a firewall on the content provider's networks (Ref Fig. 4).

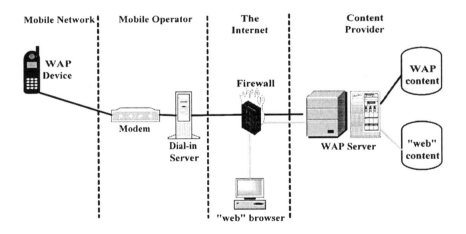

Figure 4. WAP – Functional setup – 2

The WAP device accesses Internet the same way as before, but thereafter it connects to a firewall that accepts or rejects the connection depending on the firewall configuration. The firewall passes the connection onto the WAP gateway inside the WAP server (Ref Fig. 5). In this configuration, the chain between the content server and the WAP device is secured point-to-point with the WTLS encryption protocol.

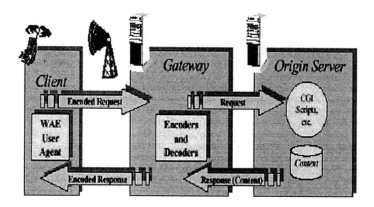

Figure 5. WAP Programming model

India, with its economical infrastructure costs, improving data pipelines, mushrooming server farms for hosting services, and availability of technically skilled manpower to provide on-going support, is becoming a major destination for hosting such services.

2.5   Bluetooth – The Smart Devices

Bluetooth represents the next level of connectivity for smart handheld devices. The development of Bluetooth began in early 1998 and was led by a number of telecommunications and computer industry leaders like Ericsson, IBM, Intel, Nokia and Toshiba, which worked to develop a global standard for wireless connectivity.

Bluetooth is an open standard for two-way, short wave radio communications between different devices.   It is a low power radio technology developed with the objective of replacing the wires currently used to connect electronic devices such as personal computers, printers and a wide variety of handheld devices such as palm top computers and mobile phones. Bluetooth operates in the 2.4GHz ISM (Industrial, Scientific, Medical) band and devices equipped with Bluetooth are capable of exchanging data at speeds up to 720kbit/s at ranges up to 10 metres. This is achieved using a transmission power of 1mW and incorporating frequency hopping to avoid interference. If the receiving device detects that the transmitting device is closer than 10 metres it automatically modifies its transmitting power to suit the range. The device also shifts to a low-power mode as soon as traffic volume becomes low or ceases altogether.

These devices are capable of linking together to form piconets, each of which can have up to 256 units, with one master and seven slaves live while the rest are in standby modes. Piconets can overlap and slaves can be shared. A form of scatternet can be established with piconets overlapping, allowing data to migrate across the networks.

Bluetooth uses a radio transmitter built into a microchip to communicate with any other device packing the same chip, thus eliminating the need for cables. Bluetooth devices create a personal area network (PAN), but unlike devices that use an infrared port for communications, direct line of sight is not needed to connect. It has higher bandwidth than current wireless communications, and data is secured by advanced encryption and authentication methods. Since Bluetooth devices transmit on the heavily used unlicensed radio band, signals sending a credit card purchase or business information competes with microwave ovens, cordless phones and other wireless networks.

In order to prevent transmissions from breaking up, Bluetooth employs frequency hopping, the practice of skipping around the radio band 1600 times each second. This improves clarity and also reduces what Bluetooth proponents call "casual eavesdropping" by allowing only synchronized devices to be able to communicate.

Each Bluetooth device has a unique address, allowing users to have some trust in the person at the other end of the transmission. Once this ID is associated with a person, by tracking the unscrambled address sent with each message, individuals can be traced and their activities easily logged. For Bluetooth devices to communicate, an initialization process uses a PIN. The security of Bluetooth is based on keeping the encryption key a secret that is shared only by participants. Since it is dependent on the other users, it definitely raises some serious questions regarding security. Authentication is used to prevent unwanted access to data and falsification of the message originator, and encryption is used to prevent eavesdropping. These two techniques along with frequency hopping and limited transmission range for a Bluetooth unit, usually 10 m, enhance the security level.

Handheld devices are expected to quickly put this technology to use but currently there are very few players in India who have made significant contributions in this area. Wipro Technologies has developed specific products and services for the Bluetooth short-range wireless standard.

3. APPLICATIONS AND SOLUTIONS – CRITICAL SUCCESS FACTORS

Currently, there are only a few companies focused strictly on providing software solutions for the mobile devices. Generally, available software is developed alongside the hardware in a closed system. As increasingly standards-based devices emerge, rapid growth in "open software" for mobile computing is expected.

Usually, the network operator provides a number of additional services to the subscriber, such as e-mail, Personal Information Manager (PIM), and a PC-based web interface that removes the constraint of a phone screen and keypad. For example, Phone.Com offers UP.Mail, UP.Organizer and UP.Web as part of its UP. Link WAP Gateway solution. The portal interface is provided for users so that they can define, organize, and personalize the services that will be accessed on the mobile device. By using a desktop PC browser, users can log on to the portal and add their favorite links to be accessed via the mobile device. A personalized menu structure can be created, and the user may decide whether data should be entered from the device or, for frequently entered information, stored securely on the portal. The Sun-Netscape Alliance has recently announced the extension of its iPlanet range of e-commerce software to include a Wireless Server for e-mail, calendar, and address book access.

The wireless Internet presents ample opportunities for applications in almost all business sectors. However, success in the new business landscape requires considering the capabilities and limitations of the technology and preparing for both. The following critical factors must be considered for an application to succeed in this emerging arena:

3.1   Strategic

Goal clarity and a sound strategy: Any high-tech company that considers wireless Internet needs to thoroughly understand the competition, the technologies that will make realization of the strategic vision and goals possible, and the major players who are potential partners and are critical in successful execution of any strategy.

Affinity of service and device analysis: Instead of rushing to deliver all possible content to all possible devices, research and analysis should be undertaken to determine the suitability of a service on the device. The vision that transcends the next generation of applications is that of device independence. Wherever it makes sense, consumers are going to demand content delivery, regardless of the devices they use: desktop (HTML, XML, XHTML), mobile phones and PDAs (WML), telephones (VXML), screen phones, televisions (ATVEF, TVHTML), pagers (SMS) etc. Since developing and maintaining separate content for specific devices is nearly impossible, companies need to move towards a model in which different templates are used to serve the same content to different devices. This will help in content reuse and better management.

Internationalization: With shrinking global boundaries, adequate attention must be paid to the internationalization issues for applications and services to be successful around the globe. This also includes rendering application in regional languages, e.g. Hindi in India.

Synergy: Cooperation and partnerships across traditional boundaries of industry and location can now create opportunities for players who offer the right mix of services. Content providers, device manufacturers, wireless carriers, and application developers depend on each other to make the most of the opportunity. The device manufacturers benefit when application developers build exciting applications that enable new services and meet new content/information demands. This in turn benefits wireless carriers, who are the benefactors of airtime usage. These partnerships not only reduce design and development costs, but also help the various players better understand consumer demands.

Leveraging the experience: Wireless Internet applications and services development follows a model similar to the wired Internet. Consequently, lessons regarding usability, personalization, information architecture, configuration options, etc., can be learned from the development of Internet applications.

3.2   Technical

Interoperability: The issues regarding interoperability of various devices, server software and hardware, and applications and services should be kept in mind since the user should not be forced to upgrade or buy a new device just to access a new site.

Emerging technologies: It is important to keep up with the specifications, to look out for technologies that the applications can leverage, and to be aware of their impact on consumers. Enhancements in other technologies will indirectly improve the applications. For example, improvement in compression techniques of graphic images (GIF to PNG) will lead to faster loading, richer graphics and a better user experience. Improvements in bandwidth technologies (ISDN to DSL to fiber optics) will allow richer interactive applications.

Security: Any breach of information is potentially more harmful in the wireless world because information would be interconnected and interdependent. Sound security procedures and algorithms are necessary at various intermediary steps.

Prototype Testing: As in any good application design, the key to any good wireless project is iteration: design, develop, learn and iterate. For successful acceptance, all wireless Internet projects need to go through the rigors of prototype testing and trials. These prototypes and trials should also address performance and scalability issues.

3.3   Market

Market analysis: A thorough, constant, competitive analysis helps companies foster growth, solutions, and success. They must keep a finger on the pulse of their potential customer base as well as that of their competitors — traditional as well as non-traditional.

Personalization: Personalized content is key to success on the Internet, and it is critical in the wireless Internet space. A well-implemented personalization strategy is essential to an effective user experience and to creating loyal customers. Content, service and application providers need to thoroughly understand the consumer before serving up information. With higher airtime costs and competitors only a click away, businesses need every possible advantage, such providing focused content, to keep the consumer from leaving

Pricing: Billing is still a challenge in the wireless Internet world. Old pricing models don't make sense in the new landscape and new options of charging need to be considered along with a new critical component of Quality of Service (QoS).

- Transaction-based billing (Per application/service/feature)
- Performance-based (Flat rate)
- Monthly fee
- Billing per use
- Or a combination of the above

3.4   Operational

Usability: Usability is important for any application or device, but it is even more important for capability-constrained devices — limited display or limited input. That gives the user a positive experience with wireless Internet applications. Usability analysis should encompass any wireless Internet project from start to finish, otherwise even the best of applications will fail to take off.

Indian software companies, with their inherent strength in developing business application, are investing significant effort to understanding the above issues and to incorporating them in their application design.

4.   INITIATIVES IN INDIA

Following table explains the Indian players and their inititatives:

Company	Contribution in Mobile Computing
Indsoft Infotech	WAP portal that features an array of services including page creation, e-mail, shopping cart, movie reservation, stock quotes, flight timings, hotel reservations etc
Integra	Developed WAP browser on Palm OS. Also unveiled the WAP server - JATAYAU. With this Integra became the 4th in the world to unveil WAP Server after Nokia, Phone.com and Ericsson.
Fascel	Internet enabled WAP. Caught up with Nokia.
Planetasia	The company is giving shape to its unique content management solution aimed at "micro-segmentation" of Internet content and services as determined by the mobile users' need and the specific requirement of telecom networks and access devices. The company's new WAP solution is first being implemented for a Kuwait based mobile portal. Micro-segmentation, as conceived by Planetasia, is expected to bolster chances of m-commerce by 50% at the individual user level. Tied up with Microland
TCS	TCS, with its global presence, is assisting Jatayau (Integra) in areas including development and test services, software engineering process and training, infrastructure, and certification. Tied up with Integra.
Visual Quest	The mobile portal (Indiawapservice.com), described as mobile window to the Internet, provides free service, information, and resources on WAP and offers services of WAP-enabling sites, building applications, training and turnkey solutions in the wireless segment. With the demand for WAP growing by the day, Visual Quest India, which has been certified a WAP developer by the WAP Forum, is also addressing training in WAP and Blue Tooth technologies.

Infosys	WAP Enabled Banking Solution: By WAP-enabling their product, Infosys has allowed banks to offer the entire range of their services to all customer segments including retail and corporate customers, on any device that supports WAP. It is both Push and Pull enabled. Thus, customers can conduct all kinds of financial activities using any device that supports WAP, including mobile phones and personal digital assistants. WAP-enabled BankAway enables banks to send critical alerts and e-mails to customers, in addition to offering routine banking functions such as account balance inquiry, transaction inquiries, de-mat account information, funds transfers, and bill payments. Furthermore, the shopping feature in the solution enables customers to make payments securely while buying goods from WAP enabled Websites. The corporate WAP access allows banks' corporate customers to monitor their corporate accounts, conduct limits inquiries and check on approvals and status of letters of credits, bills, guarantees, forward contracts, etc. The company strongly believes that WAP is going to be a key driver of mobile-commerce across the globe in the next two years. As a result, every product / service that is being e-enabled will have to be m-enabled. The company's banking product, being both WAP and SMS enabled, offers its clients m-commerce features with minimal effort.
Duncan Infotech	Wap Enabled Healthcare Applications integrated with the Hospital Management System, allows doctors to access patient records through their PDA and cell phones, check appointments, receive results of pathology tests etc. The functions related to a Nurse Station in a hospital are also made available through mobile sets.
Satyam Infoway	WAP Gateway that initially gives cricket and stock market information.
Wipro	Wipro is the one player that has made major inroads in Mobile computing with its Bluetooth solution. Wipro Technologies unveiled a suite of ready-to-use hardware and software products and services for the

	Bluetooth short-range wireless standard. The solutions are designed for high performance inter-connectivity for PCs, PDAs, mobile phones, printers, fax and wireless LAN access points. Its Bluetooth portfolio incorporates the complete Bluetooth baseband and protocol stack, as well as application profiles. Partnerships with ARM and Symbian enable Wipro to offer a wide range of services to the mobile computing and wireless markets.
Orange	It has installed a WAP gateway, which its subscribers are using to access Rediff.com's WAP-enabled content, and has introduced several SMS-based information and messaging services like mobile e-mail and mobile banking. Orange network experienced a high of 0.2 million messages being transacted daily when it offered mobile e-mail at an introductory free tariff. Although there are very few WAP phones compared to SMS capable phones, Orange is rightly playing the role of the mobile Internet incubator.
AirTel	The largest cellular service provider in India in terms of geographical spread, AirTel has tried out at least three sets of WAP solutions. AirTel has introduced unified messaging, mobile e-mail, mobile stock information services, and mobile banking. It had also experienced traffic of 0.15 million messages per day during the initial introductory phase of launching its SMS-based information services.
Aircel Digilink	It is in the process of enabling its cellular subscribers to surf WAP-enable sites. It has included Ericsson for its WAP infrastructure.
BPL Mobile	It is emerging as the most dominant cellular service provider in South India. It has also announced that it will soon introduce mobile Internet services. BPL Mobile is working on GPRS services that can significantly accelerate accessing the connection to WAP sites. It is currently in intense negotiations with at least two well-known WAP gateway vendors.

Others Cellular Providers	Other operators who are also in the process of putting up WAP infrastructure include Birla AT&T in Maharashtra, Goa, and Gujarat, Escotel in Haryana, UP (East), and Kerala, and Essar Cellphone in Delhi.
Krisn Information Technologies	It launched India's first WAP-based chat site (http://www.chatonhandy.com) and a comprehensive WAP site builder SAMRAT, which is a GUI-based WML development tool that supports all standard WML specifications like, deck, cards, table, image, anchor etc, as well as components such as POP3 Mails and Contact. It is also coming up with a WAP-to-web site converter.
Itfinity	Developed http://www.waphoria.com - the most comprehensive WAP site so far in India. It makes it easy to surf on a cell while searching for news, stocks, mutual funds, address book, bookmarks, reminders, horoscopes, movies, TV, forex rates, and currency converters.
SSKI – A broker firm	SSKI, a 75-year old broker firm with a network of 47 franchisees across 28 cities has promoted India's first WAP-enabled site about investing (http://mobile.sharekhan.com/wap/home.wml). The Sharekhan.com team is credited with installing perhaps the most advanced WAP gateway/server in India. A WAP phone can access it through any cellular service provider in the world and an ISP, regardless of whether or not the cellular operator has a WAP gateway. It is not even necessary to go through a public gateway available abroad because Sharekhan.com's gateway IP address can be used directly. Currently, stock quotes, financial news headlines, latest Sensex (BSE) and Nifty (NSE) values, top gainers and losers, and punters call, a specialized information service by stock market experts, can all be accessed.
www.rediff.com	India's leading portal, Rediff.com is one of the first Indian portals to enable access to its content through a mobile phone. Subscribers can access its WAP content through the WAP gateway installed at Orange, Mumbai. General news, business news, sports news and scores, calendar services, market indices, and daily

	horoscope/star forecasts are currently available on Rediff Mobile. URL: http://mobile.rediff.com/index.wml
www.phoneyt unes.com	This Delhi-based niche SMS and other mobile solution provider is in the process of providing a gamut of WAP-based services. It is also developing an online booking system for mobile salesman and plans to be an independent content provider.
www.netxcell. com	It is probably the first ASP in India to experiment with a mobile model. Some of the applications it hosts wirelessly are messaging, voice/fax/mail notification, m-commerce, mobile banking, corporate e-mail, and personalization. Netxcell is in the process of launching its site and is likely to brand it with a carrier's name. Most services are planned to be free offerings.

5.   ADVANCES AND FUTURE TRENDS

Mobile technology is no longer the domain of executives on the move. The emergence of competitively priced mobile phone technology with the capability to address the Internet is an evolutionary development and has made the handset a more common tool in the common man's hands. According to NASSCOM, India has a mobile population of 2.88 million and of the total Internet users around 20% own credit cards and around 14% own mobile phones.

In order to tap this market, the solutions need to be innovatively designed, keeping the specific needs of the Indian consumer in view. Some of the trends and the norms that are foreseen in the mobile computing world are:

Mobile enterprise applications: Companies can reap huge savings when they make the information available at employees' fingertips. With improvements in processing power and network bandwidths, enterprise applications will be extended to handhelds and mobile devices. Gartner Group predicts that by 2004, 80% of new applications for mobile/consumer use will permit access from mobile phone clients. Wireless e-mail is the first logical step to establishing essential mobile systems in an enterprise.

Interface to mobile agents and e-services: Mobile agents and e-services adhere to the concept of applications and services automatically communicating with each other

Next-generation wireless networks and devices: 2.5G/3G wireless networks promise increased reliability and extended bandwidths (up to 2Mb/s transmission rates), which will help bolster consumer demand, thus enabling innovative multimedia services and applications.

WAP-enabled intranets/extranets: As corporations and businesses realize the power of wireless Internet, corporate intranets and extranets will also be extended to serve a variety of wireless devices.

Advanced personalization: As application and content providers improve their understanding of consumer needs, advances in personalization technologies will help pull and push customized information and manage content delivery across various access devices different in shape and form.

Location-based services: Once the locations of a device and the consumer are known, the horizon for value-added services and applications is endless. What truly makes m-commerce different from simple e-commerce is the ability to sell to users based on their location.

Personal portals: The lines between personal and professional data will continue to blur. Consumers will have the option of combining and configuring mobile phone/PDA menu items based on their personal needs and preferences.

Voice recognition boom: Gartner Group predicted that by 2001, voice interface will become an expected option in retrieving Internet-based content. The world is in the early development phases of voice-based portals and Web sites that take advantage of text-to-speech technologies.

Biometrics: Biometrics (retina-, voice-, face-, and palm-print based authentication) and Smart Card integration into mobile devices will improve security, thus enabling mobile e-commerce and launching new applications.

6.   CONCLUSION

We are experiencing first hand the differences that can exist between a declared standard, such as WAP, and the emerging standards that the Internet produces each year. Although the existing applications and possibilities in the wireless Internet marketplace are exciting, the advances and future trends listed above will further define the economy and, in turn, our society. In India, the future has never looked as high-tech as today.

The Indian web and Internet companies simply cannot risk not going mobile. The most accepted parameter to judge the value of a website is the number of eyeballs that visit it and the time spent surfing, whether it is a content company, or the volume of commerce transacted out of the site for an e-commerce company. In a country where Internet subscriptions through telephone lines are still limited and PC and telephone penetration is not very good, the question is, can a company overlook targeting an alternative audience that is being opened up by mobile phones?

The cellular industry, which still now had based its business plans on proven models of revenue streams until now, is suddenly faced with making business plans based on much more tentative parameters—just like the Internet services/dotcom company. Grabbing the eyeballs is a new practice that the mobile operators are gearing up for. Nobody knows whether these eyeballs will convert into revenue. A mobile service company, like the traditional company, may say "no" to this way of doing business. However, if there is a huge business opportunity based upon Internet technologies in the near future, it then risks losing out on that opportunity. In spite of the dotcom meltdown, the experts still see an Internet economy emerging. If the pundits are to be believed, then the mobile operators cannot afford to forgo the mobile Internet way. In any case, the investment on WAP gateway is not much to be concerned with when millions of dollars have already been spent on acquiring licenses for setting up mobile networks. Furthermore, unlike the Internet companies, the mobile operator already have cash flowing in through other revenue streams such as airtime tariff and SMS tariff. For solution providers to the emerging wireless data, Indian mobile operators and WAP service providers are attractive customers with whom home business relations can be developed for greater future potential. As mobile phones with greater capabilities enter the market, the scope for services like wireless Internet will expand. It is then that India will emerge as one of the biggest lands of opportunity.

Indian companies are also looking keenly at the export market for WAP products and services. In fact, some of the companies here believe that currently the export market that should be focused on more aggressively. WAP is almost as new to the world as it is to India. There are so many countries, more economically developed than India, where the stage of WAP deployment is not as advanced as it is in India. This scenario has already bred several Indian start-ups, which are focusing on developing solution and service skills to target the wireless market. In fact, some of the early birds like e-Capital and Planetasia have had successes abroad. e-Capital has done impressive projects for big clients like WapIT, Yomi Media, and Sonera in Finland, and Vodafone in the UK.

Competition between mobile service operators is rising enough to create a serious need for innovative value-added services through mobile phones, with easy and inexpensive access to applications. Only these moves would attract new customers and boost revenues. The ability to launch new services on a continual basis will be critical for the survival of mobile operators in the future. This is one factor that will ensure the latest technology remains within your grasp.

REFERENCES

[1] *Report-Internet Service Providers Association of India*, ISPAI.
[2] C. Arehart et al, *Professional WAP*, Mass Market Paperback, Wrox Press Inc, July 2000.
[3] M.V.D. Heijden and M. Taylor, *Understanding WAP Wireless Applications, Devices and Services*, Artech House, July 2000.
[4] Andy Dornan, *Essential Guide to Wireless Communications Applications: From Cellular Systems to WAP and M-Commerce*, Prentice-Hall Computer Books, November 2000.
[5] http://www.phone.com
[6] http://www.wapforum.com
[7] http://www.indiamarkets.com
[8] http://www.indiainfoline.com
[9] http://www.wap.com

APPENDIX –1: CASE STUDIES

WAP-based Mobile Banking: UTI Bank Ltd.

The WAP based mobile banking services will provide information such as online balance updates and details of the last five transactions, in addition to taking requests for checkbooks and statements. The solution, which will initially only provide information, will offer personal financial services, including personal banking and on-line information on utility bills, in a phased manner.

Technology (short service messaging) works on e-mail. As part of its computerization drive, the bank is migrating from a decentralized database to a centralized one. Although only four branches have been linked so far, the bank plans to link all its branches.

The centralized database program is being connected through leased lines, with V-SAT back-up. The bank has budgeted an amount of Rs. 32 Crores for its computerization program, including Net banking. The bank will launch Net broking soon. UTI Bank will also make a decision on setting up a payment gateway shortly. For its ATM program, the bank has joined with MasterCard and plans to join with Visa International.

Mobile Banking: ICICI

ICICI has launched WAP-based mobile banking services for its customers. The banking services include online balance update, check book request, details of last five transactions, statement request and verification of TIFD status in the DEMAT account. This solution will also enable an ICICI credit card customer to check details on his outstanding balance, payment status, and cash and credit limits in the last three transactions.

Mobile banking using WAP technology allows secure online access of the Web using mobile devices instead of SMS (short messaging service) technology, which works on e-mail. It launched real-time stock quotes and online monitoring of the performance of equity portfolio and also introduced ATM and branch locator facility for WAP phone users.

ICICI proposes to launch personal financial services on WAP, including personal banking services, online information tracking of utility bills, and travel and ticketing information.

WAP-enabled Banking: IDBI-Bank

IDBI is initially providing its WAP-enabled services in Mumbai, where it will allow its customers to receive balance, details about recent transactions, check paid status, and stop payment status, as well as request for checkbook and account statements.

The bank plans to add ATM locators and fund transfer facility along with E-Sec accounts, and SMS. According to IDBI Bank's managing director and

CEO Gunit Chadha, the bank is initially targeting its customers with mobile phones.

Mobile Surfing: Indiatimes – Hutchison

INDIATIMES.COM joined with çellular service provider Hutchison to offer news and information to WAP phone users in Delhi and Mumbai. Hutchison subscribers will get the latest news from Indiatimes.

Orange subscribers in Mumbai and Essar subscribers in Delhi will get access to round-the-clock weather updates, stock quotes and forex rates, astrological predictions, links to various WAP sites, and STD codes of major Indian cities.

Hotel guides for Ahmedabad, Bangalore, Calcutta, Chandigarh, Chennai, Delhi, Jaipur, Lucknow, Mumbai will be just a click away. Services in the offering include contact numbers and addresses of police stations, hospitals, chemists, blood banks and fire brigade services for eight major Indian cities, and airline and railway schedules. Online stock trading, movie ticket sales, and the Google Search to cut through the clutter on the Net are also being planned.

Mobile Surfing: Tata Cellular

Hyderabad-based Tata Cellular is the first Indian cellular service provider to enable its subscribers to surf the Net through a mobile phone. In the first few days of mobile Internet services, this operator had introduced mobile Internet services with free airtime.

Even later on, the fee was still only Rs 2 per minute for surfing WAP content, which is significantly less than the Rs 4.5 per minute standard airtime tariff. It has also allowed its subscribers to surf any WAP-enabled site in the world through its WAP gateway.

Aside from SMS-based e-mail services, the company has yet to go in for other SMS-based services like stocks@tatacell.net and mobile banking. Introducing these SMS-based services is a priority because of the huge number of SMS-capable mobiles already existing in the country. However, it was the long-term opportunity of convergence technologies and the first mover advantage was what really pushed Tata Cellular to enable Internet access on its mobile network. It has already invested Rs 10 million on WAP and expects the ongoing roll out of mobile Internet to take less than 6 months.

Chapter 2

COORDINATING TECHNOLOGIES FOR VIRTUAL ORGANIZATION

Joyce Lucca, Ramesh Sharda, Mark Weiser
Oklahoma State University

Abstract: In order for organizations to succeed, they must be able to respond with flexibility in a geographically dispersed environment. Virtual organizations, which can form, disband, and re-form to meet ill-defined and emerging situations are playing an important role in organizational strategies. For virtual organizations to be successful, system level elements such as organizational memory, hardware, and software must be identified. Also a number of knowledge management tools such as database access, expert systems, and groupware must be in place. Another critical enabling component to this scenario is network technology, which physically connects the various modules. This is a multi faceted issue. This chapter addresses the need for a cross-disciplinary approach to bring the pieces together. A comprehensive review of current literature in these related areas is given. Also, a framework for linking relevant technologies and a closer inspection of the database segment are presented. The chapter concludes with a description of such a framework in place in a virtual organization, and research issues in this area.

Keywords: Enterprise resource planning, Globalization, Personal digital assistant (PDA), Virtual organization, Virtual management enterprise

1.   INTRODUCTION

Pervasive computing provides an ability to access information from virtually anywhere at any time, potentially transforming the way we live and work. Police Officers can retrieve critical information from their cars, airline customer can get updated flight schedules from their Personal Digital Assistant (PDA)'s, and shop floor computers can access critical information to ensure efficient production workflow. Increasing availability of computing resources and data to portable devices and different environments has great potential to alter our business and personal lives.

The fusion of computers and communications allows for improved communication and efficiency in the work place as well as distribution of work globally to take advantage of distant expertise. This is leading to a paradigm shift in organizational structure known as the virtual organization. Virtual organizations, which can form, evolve, disband, and re-form to meet ill-defined and emerging situations are playing an important role in organizational strategies. Virtual organizations face even greater challenges in coordinating their activities and managing the knowledge spread across the team members than their traditional counterparts.

The term "pervasive" may be taken more broadly to include the critical nature of computing to many different functional areas, in addition to the increased accessibility that enables the infrastructure. The reach of computing has altered the internal function of some organizations, and has even created new businesses and business methodologies. A virtual organization is one such new entity. The concept of virtual organizations has been around for quite some time, but only with the relatively recent improvements in the computing environment are they widely viable. An expanding array of tools is available to aid knowledge workers in a virtual environment. These include such things as both wired and wireless telecommunications technologies, object oriented software designs, Enterprise Resource Planning (ERP) integration, and robotics. This chapter focuses on the technologies necessary to support a virtual structure and provides a system framework for the coordinating technologies of virtual organizations including current knowledge management tools. A conceptual coordination architecture for a virtual manufacturing organization, including a comprehensive description and model for data repository linkages is discussed. A brief description of the partial implementation of this architecture in an actual virtual organization is also presented.

2.   THE EMERGENCE OF VIRTUAL ORGANIZATIONS

The 1990s saw an explosion in innovation and technology with migration toward a knowledge and service-based economy. Integration of work processes with electronic network infrastructure was seen to be rapidly changing the dynamics of organizations. The virtual organization structure is emerging as a means for enabling companies to stay viable. Virtual organizations blur the lines of hierarchy and can form, evolve, disband, and re-form to take advantage of core competencies to meet ill-defined and emerging situations. Donlon [1] defines the virtual organization in terms of core competencies. "The virtual organization is a bundle of competencies, some internal, some external, arrived at through relationships with other people and pulled together to deliver a value". Hedberg et al [2] contend that companies which identify and develop core competencies will be able to maintain a strong position when partnering with outside firms. They identify core competency with several key questions: i) Does the product or service differentiate the organization relative to competitors? ii) Does the product or service give the organization a distinctive advantage? iii) Does the product or service permit different applications? iv) Can the product or service be sheltered from imitation by the competition? Sometimes the non-core activity of one organization is the core strength of another's and therefore the synergistic strengths can stimulate the growth of alliances [1]. These relationships are based on competencies, not historical relationships, to bring the highest quality product or service to market as quickly as possible [3].

Advantages of virtual organizations include flexibility and globalization [4]. With respect to globalization, Boudreau [5] points out that virtual organizations are a collection of federated enterprises through arrangements such as joint ventures, strategic alliances, consortia, coalitions, and franchises, and cites the sports footwear industry as a good example. Almost all of Nike's and Reebok's production is outsourced to Asian countries. Virtual organizations such as these can operate independent of time and space by connecting geographically dispersed resources. Flexibility is another important advantage because opportunities, especially in the global market, continually shift. Companies that cannot respond quickly may end up with under-utilized resources. Another benefit is the ability to tap into the much richer innovation skills that outsourcing can provide [6]. Additional advantages include improved resource utilization, and lower costs [7]. Virtual organizations allow smaller manufacturers to pool their core-strengths and compete for projects for which they would not otherwise be able to bid.

Disadvantages of the virtual organization include possible loss of process control and the fact that proprietary information must be shared, requiring a high degree of trust among participants [4]. Another disadvantage is that extensive use of switching of employees or suppliers may have a negative impact on the virtual organization [7]. This points to the need for the virtual organization to create the appearance of consistency, and strive for the seamless integration of switching activities.

Whereas virtual organizations can sometimes solve problems quickly and easily, they typically make decisions in unstructured environments under tentative conditions. This requires dispersed virtual teams to develop an organized and well-structured knowledge management system to bring people together to distribute and enhance the expertise of the group.

The coordinating entity is the store of organizational memory as well as the full context of this specific virtual group. The coordinating entity brings all the virtual enterprise members together to be able to compete for business, then design, develop and deliver the product/service on time, and process payments to/from all the enterprise participants. It is obvious that a considerable amount of information exchange is necessary to make it all work well. This collective knowledge store is managed, processed, and used through a whole host of coordinating technologies described below.

3.   COORDINATING TECHNOLOGIES

Coordinating technologies can support the various needs of knowledge workers in virtual organizations. These collaborative technologies can help virtual organizations to maintain and re-use knowledge, and to enable team members to acquire the tacit knowledge of other team members. We can understand the coordination system in a virtual organization along the traditional modes of analyzing hardware, software, processes, and people. Each of these components is discussed below.

3.1   Hardware

Computers continue to develop by becoming faster, more powerful, and less expensive. Communications technologies continue to improve providing more and faster channel capacity. The combination of these improved technologies is important to the virtual organization.

In terms of hardware, the most important component is the personal computer or workstation [8]. With the proliferation of personal computers no special desktop hardware is needed other than sufficient memory and processing speed capable of supporting groupware technology. The most common configuration for a networked organization is client-server based and is more efficient for all but the smallest networks [9]. This allows network resources to be concentrated on a few highly powerful servers and requires little configuration on the part of the client. This methodology of using low cost network computers to reduce expenses and complexity is known as thin client technology [10]. One major advantage of a thin client platform is that application upgrades or changes are much easier to implement because the application only needs to be updated on the server. In a distributed environment, an open architecture that ensures interoperability is an important consideration [11]. The ability to interchange components for both the client and server, allowing for flexibility is a very desirable feature.

The hardware for communication channels depends on the network. In the past organizations relied on voice channels that were also used for telephone conversations to transmit data [8]. However, with the movement toward the use of multi media technologies, voice channel capacity is being surpassed. For media rich applications, organizations are turning to high-speed technologies that can efficiently handle the combination of voice, video, and data traffic simultaneously.

The choice between public and private network technology is largely a function of the size of the organization. The most widely used public network is the Internet, which physically links computers across the world. The obvious advantage of the Internet is reasonable cost, with the disadvantages being unpredictable speed and lack of security. Choices in private network technology include intranets and extranets. An intranet is a network within an organization that uses familiar browser technology for collaborative work within the organization and basically provides the same services within the organization as the Internet provides between organizations [8,10]. Extranets are an extension of Intranets to customers, suppliers, and other team mates. Access is restricted to only authorized users through password protection. A logistical, robust network can be a key factor in success providing better information, faster speed, and more coordinated distribution [12].

3.2   Software and the Business Process

The umbrella term that is often used to describe the concept of collaborative technology
    is computer supported cooperative work (CSCW). This developing field of study involves the interaction of computer technology and people working in groups in a distributed environment. From this idea, software has emerged to enhance cooperation, communication, and coordination and is known as groupware. Groupware applications help individuals to share information and coordinate work activities [13].

One of the earliest, simplest, and most successful groupware tools is electronic mail [10]. It is fast for both sender and receiver and for both review and reply, it is easy to use and understand, and has wide-cross platform support [14]. Because speed is an important element in the product cycle process, electronic mail has been an extremely popular and helpful tool.

A more sophisticated groupware application is conferencing. Computer based conferencing is rapidly becoming an essential component to the operation of a virtual organization [10].

Conferencing can be either text based or video based. Text based conferencing is based on the notion of a discussion thread where different group members discuss a particular topic. Users can create new threads or reply to existing posted discussion threads.

In contrast, video conferencing is typically synchronous in nature and provides a more media rich environment for the users. Recently desktop video conferencing has greatly increased in popularity. There are a number of advantages to desktop video conferencing: i) when a meeting needs to be scheduled, it is not necessary to find a mutually acceptable location, ii) no traveling is required, iii) physical expressions and body language can be conveyed, and iv) physical objects such as products can be seen by all participants [10,15]. In addition, most video conferencing products also include shared application functionality such as whiteboards that allow for multiple authoring of documents.

There is also project management groupware used to facilitate scheduling and calendaring to expedite group coordination [10,13]. For example, when a group member needs to schedule a meeting, the application is capable of examining other members' schedules to locate an optimal time, notifying members of the meeting and negotiating meeting conflicts based on name or position within the group.

One of the most sophisticated applications to enhance group performance is expert or knowledge base systems. These systems are used to capture the expertise of exceptional employees and codify it in a format so that others can use it collectively in the organization [16]. The system consists of a knowledge acquisition system, an expert database, and an inference engine. The knowledge acquisition facility is the software interface that creates a dialogue between the system and the human expert in order to acquire human knowledge.

3.3   Human Factors

Groupware modifies the ways in which people interrelate. When new technologies and change are introduced into an organization, human factors must be considered. When implementing change it is critical to have the support and dedication of key stakeholders who can recognize economic incentives that provide good reasons for implementing change [17]. Markus et al [18] address the issue of change with respect to information technology. The authors argue that information technology is not a magic bullet but that behavioral flexibility is the key factor in a successful change process.

Organizational culture is also an important consideration. Introducing new technologies such as electronic collaboration and knowledge sharing programs without having the proper organizational culture in place increases the likelihood that the full benefit of the investment will not be realized [15]. The author cites three important factors: i) when rewards are based on individual performance, people are less likely to participate in collaborative activities, ii) organization conflicts will have a negative impact on the use of collaborative technology with those viewed as rivals, iii) without some type of coaching or training, people have a tendency to use new technology in the same way as the previous technology.

Another related issue is trust. Coleman [17] cites a study involving a virtual organization where some US team members stayed in the homes of their Swedish teammates and other US team members stayed in a hotel. It was found that the team members that were involved in a "home stay" developed a much higher level of collaboration. For the US members, having contextual insight into their hosts' lives resulted in a higher level of trust, enabling improved communication.

4.   AN OVERVIEW OF PROCESSES IN VIRTUAL MANUFACTURING ORGANIZATIONS

Having provided an overview of the coordinating technologies necessary for a virtual organization to function, in this section we present a suitable process model and provide an in-depth look at communication linkages for a Virtual Manufacturing Enterprise (VME) organization. Then, we posit some applicable coordination technologies to support these processes.

For our model we assume a manufacturing environment requiring a bidding and award process, such as that required for many government or industrial contracts. A basic retail model would only simplify the model by eliminating some of the earlier costing steps. After describing each process in this VME environment, we propose applications of some coordinating technologies to support the specific process.

As the name implies, the VME will not likely do any of the physical manufacturing, but will instead coordinate the activities of organizations that have excess capacity and the capabilities to perform some or all of the required process that comprise the desired product. The virtual enterprise needs to examine its IT requirements to support the planning and control tasks of a production system. We can consider three major external entities dealing with the VME process: the potential purchaser, cooperating manufacturing (CM) entities, and raw materials suppliers (Figure 1).

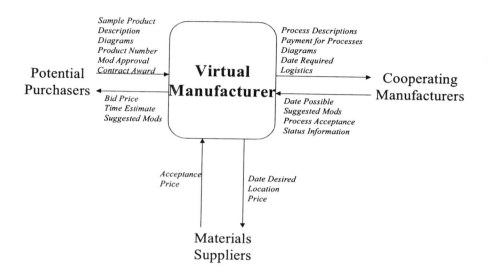

Figure 1. VME components

The potential purchaser must provide detailed specifications for the item that they would like to acquire. This may include written descriptions, engineering diagrams, a well-known part number, or even a physical sample of the part. The VME must then estimate the manufacturing cost of the product, based on engineering expertise, prior experience, and any special requirements. After a profit margin is added to the basic cost, a bid is submitted with an estimated completion time. Additionally, modifications to the requested product may be suggested to the potential purchaser. These are usually due to advances in manufacturing processes or superior materials that were not available when the product was originally designed.

If the purchaser accepts the proposal of the VME, the work must then be assigned to Cooperating Manufacturers (CMs) for completion. This assignment is based upon mutual agreement of price and timing, and an assessment of response to prior contracts. Raw materials are then supplied by a vendor selected on the basis of ability to meet target dates and proximity to CMs.

The processes that are used to appropriately bid on manufacturing jobs that will be manufactured by other organizations and those for determining appropriate manufacturers who can participate in the production of a single product determines the viability of virtual manufacturing. Rapid decision-making in a geographically and organizationally distributed environment are critical elements in this process. Because multiple organizations will ultimately be involved in a single contract, knowledge of engineering, processes, and corporate "citizenship" within the virtual manufacturing environment must be maintained by the VME. The VME will enter into contractual agreements to supply products to the consumer and to have processes performed that will create the products.

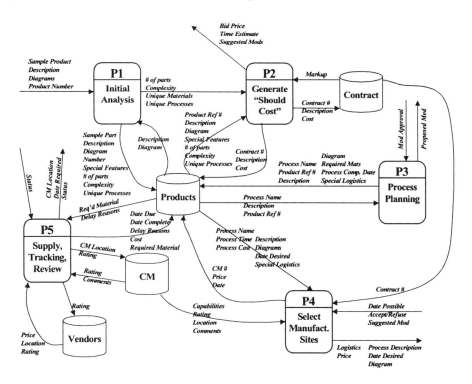

Figure 2. Major processes

Figure 2 describes the major processes for coordination of a virtual enterprise. When an opportunity to manufacture a product is received by the VME, an *initial analysis* is performed to determine the cost drivers for the product and any special requirements (P1). Based upon that information and similar past products, a bid is generated, along with an estimate of production time and any suggested modifications to the original request or design (P2). If the contract is awarded to the VME, a more detailed analysis is performed to better determine the component processes and materials that will be necessary to create the desired final product (P3). Using a list of cooperating manufacturers who are capable of performing one or more of the processes, a combination of CM's are selected, based upon past performance, proximity of other CM's, and capability to meet the timing requirements (P4). A raw materials vendor is then selected whom can best provide for the selected manufacturers. The process is then tracked from supply, through manufacturing, and delivery (P5).

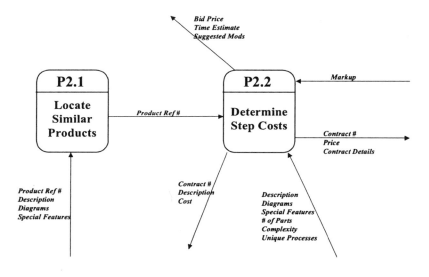

Figure 3. Initial analysis

4.1   Initial Analysis

Initial analysis may be as simple as matching a well-known product number with stored data from when the VME previously produced the same item (Figure 3). Information on the past processes, costs, and unexpected problems can be retrieved. Otherwise, an engineer must determine the number of component parts and processes, manufacturing complexity, and unique materials and/or processes that will be required. The detail of this information is dependent upon the complexity of the part and the engineer's experience with prior similar product production.

At this point in the process, communication is mainly internal, using a VME based intranet. The appropriate groupware to enhance internal communication would be email and text based conferencing. The silo of information most important to this process is Products which contains information about past production of the same or similar parts.

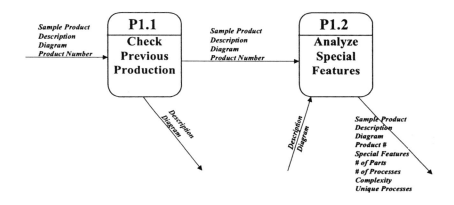

Figure 4. Cost estimation

4.2   Cost Estimation

Using this information and actual production costs of the same or a similar product that has already been produced, an estimate of what the completed product *should cost* is computed (Figure 4). Cost of materials, production, and special requirements of the contract, and shipping and other unique logistics are used to compute a cost. Depending on the class of part (i.e. electronic, steel fitting, armament, etc.), a standard mark-up is applied. Based on that information, a bid for the VME to produce the product is submitted to the potential purchaser. This figure usually must be determined quickly in order to qualify to bid, however, accuracy is important to ensure a profit for the VME, while meeting the requirements of the contracting agency. Most parts created by virtual manufacturers are not "commodity" items, but are small quantities of highly specialized products. Often a pricing decision must be made for a product that a VME has never constructed before. Experience, "rules of thumb," and reference to similar parts processes all feed into the generation of the estimate.

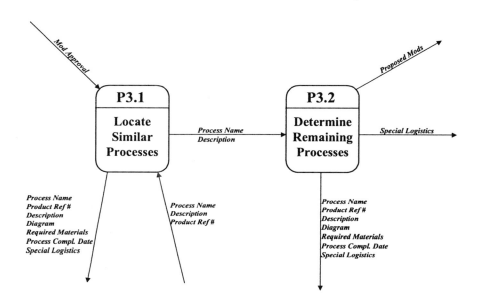

Figure 5. Process planning

Communication is still mainly internal with intranet, email, and text-based conferencing providing the most functionality. The database silos most important to this process are Products, Vendors (which includes pricing information), and Logistics (which includes information such as cost of transportation). An expert system would be helpful in re-using previous "rules of thumb".

4.3   Process Planning

If the contract is awarded, a more detailed analysis and *process planning* must be done in order to divide the work and arrange for the physical manufacturing (Figure 5). An engineer will build on the initial analysis report and specify the logical steps required to turn raw materials into the finished product. To simplify the task, similar processes that have been performed on other products are located, along with costs and any special problems. For the remainder of the processes, any specialized tests, such as chemical analyses or metallurgy are done to ensure compliance in the final product. Requirements for each of these processes and the facility that will complete them are tightly defined. Any specialized transportation requirements, due to the nature of the product are noted, and passed on for assignment to individual manufacturers. Each process, including specialized transportation, is then assigned a cost.

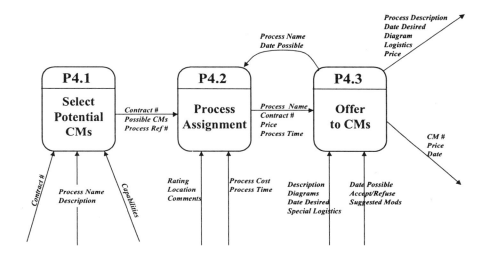

Figure 6. Task assignment

Communication is still primarily within the organization so intranet, email and text based conferencing are effective technologies. The database silos most used in this process are Products and Logistics. At this point, project management software would be helpful for coordination and the scheduling of meetings. An expert system would add value here to enable the re-use of data to provide best-practices-type information.

4.4 Task Assignment

The VME must maintain relationships with a range of CMs (establishment of these relationships are not addressed in this chapter), to allow them to quickly assign processes to physical manufacturers (Figure 6). Based upon collected information of CM capabilities, all manufacturers that have the potential to complete any of the component processes are identified. The combination of processes that must be performed, geographic locations of potential CMs, and prior experience with different companies are matched with suitable manufacturers. These manufacturers are offered the chance to complete specific tasks at an assigned price, and may accept, decline, or counter-offer. Further analysis is done to locate secondary manufacturers to complete processes that are declined in earlier offers.

Communication outside the VME begins to take place at this point. For security purposes, Value Added Networks (VANs) and extranets would be beneficial. Of course, internal communication still continues using the intranet. Email and text based conferencing still provide functionality, but video conferencing would add an element of media richness. The CM could view actual physical parts and actual shop floor views could be seen by the VME. All four silos of the database are used in this process. An expert system would aid in the decision process for CM selection, providing information on past history, including quality of work, adherence to scheduling, etc.

4.5   Control-Tracking

The fifth process actually represents three loosely related processes (Figure 7). While manufacturers are being located and contracted to do the processing, materials suppliers are identified (P5.1). Specialized materials may have sole suppliers and/or may need to be back-ordered or specially produced. Ideally, all required raw materials would be available as soon as the CM's are prepared to start the manufacturing process. Similar to selection of manufacturers, past experience with competing vendors and their proximity to selected manufacturers may impact which vendor is selected.

Tracking (P5.2) and Assessment (P5.3) are potentially the most critical of the virtual manufacturing environment -- even more critical than on a traditional shop floor. Not only can a close scrutiny of an on-going process maximize the chance for meeting contract requirements, but also reassessment of both vendors and manufacturers will lead to continuous improvement of the overall virtual environment. For instance, if a certain machine is down in a physical shop, the manager can quickly reallocate the work to other appropriate equipment or estimate a delay if alternative equipment is not available. Unfortunately, the VME does not have this immediacy to the process, however, it is the VME that is responsible for satisfaction of the contract to the purchaser.

There must be an appropriate feedback loop from each CM to the VME, indicating processing status. At a minimum, this should be at the granularity of the processes determined by *process planning*. If a specific identifiable process will not be completed on schedule and the VME knows about it, including the expected delay and the reason (equipment failure, scrapped part, delayed raw material delivery, etc.), they will better be able to manage the entire process. This may include simply notifying other subsequent manufacturers or re-allocating work to a different CM, since the virtual shop floor includes many physical shop floors.

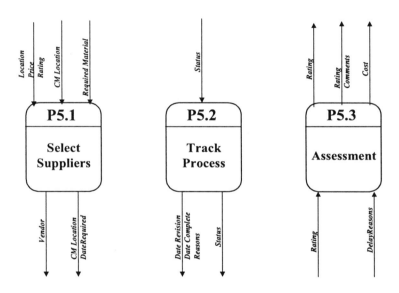

Figure 7. Control tracking

Communication in this phase is also primarily external, with VANs and extranets providing extra security, and video conferencing enhancing the richness of communication. As always, intranets and text based communications play an important role. All silos of the database are needed in this phase as well.

Table 1. Product and process summary

	Initial Analysis P1	Cost P2	Process Plan P3	Assign P4	Track P5
Hardware					
VAN				X	X
Intranet	X	X	X	X	X
Extranet				X	X
Software/Groupware					
Email	X	X	X	X	X
Text Conf.	X	X	X	X	X
Video Conf.				X	X
Project Management			X		
Expert Systems		X	X	X	
Database					
Products	X	X	X	X	X
Coop Mfg.				X	X
Vendors		X		X	X
Logistics		X	X	X	

4.6   Process Summary

In the above discussion, we have illustrated that the VME can be segmented into five distinct processes with key coordinating technologies identified for each process. A combination of all the technologies facilitates a wide range of options that enable the VME to have a high level of flexibility and functionality. Whereas certain technologies such as intranets, email, and database product information are useful across all functional processes, others such as VANs, extranets and videoconferencing might not be necessary until the later processes are performed. For efficient project management and completion in the final stages, increased emphasis on technology is required. These ideas are illustrated in Table 1.

5. A PROTOTYPE IMPLEMENTATION IN A VIRTUAL ORGANIZATION

This section describes one implementation of coordinating technologies in a virtual organization. Our focus is on VMEs that are being established to resolve the problem of part obsolescence in aging weapons systems. This has been accomplished through a dynamic supply chain and has supported various requirements for both the US Air Force and Navy. Specific products include components for landing gear, missiles, laser-guided munitions, aircraft structures, automated test equipment, and ground support equipment. The VME approach has provided substantial cost savings to the government as well as significantly reducing lead times. VME operates as a virtual machine shop spread across different organizations throughout the US. It incorporates many of the elements of enterprise resource planning (ERP) and manufacturing resource planning (MRP) and is capable of using the excess capacity of member VMEs.

An existing, functioning VME is Small Business Innovation Research Engineering Companies, Inc. (SBIRE), a small business that provides On-Demand Manufacturing (ODM). SBIRE's core competency is the capability to support critical, reverse engineering requirements for both performance optimizing programs and in the manufacture of out-of- production parts and assemblies by leveraging the idle time and machinery of its virtual enterprise partners which are companies or organizations who are committed to work in some facet with the SBIRE. There are three types of partners: i) industry, ii) academia, and iii) government. SBIRE has partners throughout the US. The primary customer base for SBIRE includes prime contractors of weapons systems to the US government such as Boeing and McDonnell Douglas, and government entities such as Tinker Air Force Base.

SBIRE is currently working on several important projects. The US Navy is involved in a multi-phase process to reduce problems with obsolete spare parts. It is working with SBIRE for the re-manufacture of over 20,000 ammunition containers. Parts are being stored in the Army Ammunition Depot in McAlester, Oklahoma and are transported to SBIRE on an as needed bases. This has resulted in significant savings in transportation costs. SBIRE is also in the qualification process for multiple configurations of C-130 aircraft control cables. SBIRE has acquired all the necessary documentation and has received an initial engineering approval for seven different configurations from the Air Force. Upon final approval, SBIRE and Ogden Air Logistics Center will enter into a contractual agreement. Using just-in-time manufacturing principles, SBIRE will deliver the cables within seven days after notification.

A proprietary database, logically similar to the descriptions in Section 5, is the central component in the operation of SBIRE. Members are linked through the Internet using a common advanced object management system known as The Object Czar ™ (Oz). Oz has a "Windows Explorer"-like graphical user interface (GUI) that creates and maintains hierarchically structured, object-oriented database applications. Oz was developed for use by government, commercial, and academic organizations to aid in the rapid development of business, engineering, manufacturing, and scientific applications that require a hierarchical class structure (Figure 8).

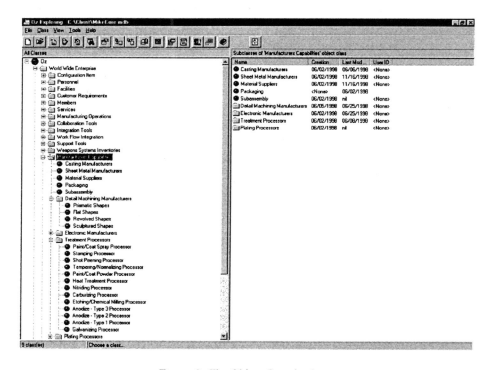

Figure 8. The Object Czar database

Oz allows managers to have instant access to mission-critical information, enabling accurate cost estimations. Oz is also capable of generating important planning data such as Gantt charts, graphs, and other comprehensive reports. Oz allows project team members throughout the world to share data including part specifications and drawings. Oz has the functionality to track and flag important things such as cost and time overruns with "stoplight" color changes at all levels of the project, allowing for efficient project management. Security issues are managed by Oz in that support for multiple users is handled by each having different privileges to Oz capabilities and class levels (Figure 9).

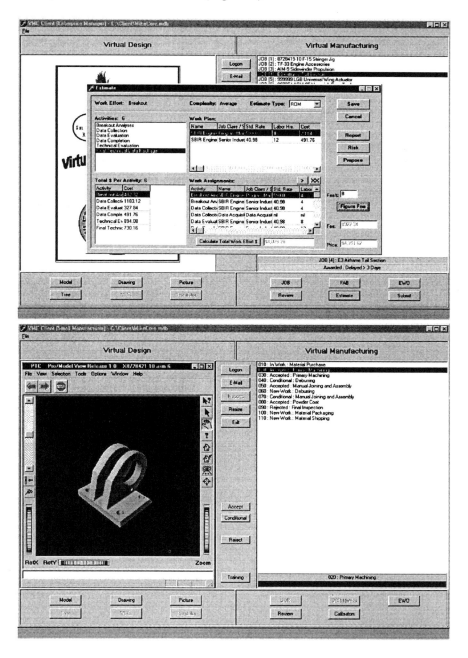

Figure 9. Views from within OZ

6.   CONCLUSIONS

The constructs of virtual organizations potentially allow small organizations, which could not otherwise handle complex manufacturing tasks that require precision in each subtask, to specialize in a component of the overall process. By teaming with other entities that specialize in different manufacturing processes, the breadth of the complex tasks can be accomplished, allowing more companies to compete in large bids. This provides small manufacturers with the potential for greater growth and profitability. A twofold benefit can be realized by the purchasing organization in more competitive pricing, and improved quality.

Collaboration does not occur because people are connected electronically. Virtual organizations and their coordinating technologies are providing a framework for an understanding of how best to work in the Information Age. Timely information is critical to a virtual organization. Pervasive computing is rapidly emerging as a means to reduce the chances for failure by enabling both the push and pull of information anywhere at any time. It is important that these technologies be thoroughly explored.

Control and tracking the progress of industrial orders being manufactured by a virtual organization is more readily possible with pervasive computing devices. A cross-disciplinary perspective combined with careful examination of the necessary components of a successful information technology system may provide the basis for advances in virtual organization development. Through our investigation, we found the need for future study to provide a more detailed analysis of the database and the linkages to it. Other possible related research streams include developing applications for emerging technologies such as data warehousing and expert systems, as well as evaluation of groupware products. Also, security issues are arising as traditional boundaries are overlapping and the need for access of team members to some, but not all, information is becoming an important concern that needs to be studied more carefully. Effectiveness of these technologies and related human factor issues are yet other areas for further investigation.

REFERENCES

[1]   J.P. Donlon, "The virtual organization," *Chief Executive*, 1997, 125: pp. 58-66.
[2]   B. Hedberg, et al., *Virtual Organizations and Beyond*, West Sussex, England: John Wily & Sons Ltd, 1994.
[3]   R. Grenier and G. Metes, *Going Virtual: Moving Your Organization into the 21st Century*, 1995.

[4] T.J. Strader, F.-R. Lin and M.J. Shaw, "Information infrastructure for electronic virtual organization management," *Decision Support Systems*, 1998, 23: p.. 75-94.

[5] M.-C. Boudreau, *et al.*, "Going global: Using information technology to advance the competitiveness of the virtual transnational organization," *The Academy of Management Executive*, 1998, 12(4): pp. 120-128.

[6] J.B. Quinn, "Strategic Outsourcing: Leveraging knowledge capabilities," *Sloan Management Review*, 1999, 40(4): pp. 9-21.

[7] A. Mowshowitz, "Virtual Organization," *Communications of the ACM*, 1997, 40(9): pp. 30-37.

[8] I. Hawryszkiewycz, *Designing the Networked Enterprise,* Norwood, MA, Artech House, 1997.

[9] J. Casad and D. Newland, *MCSE Training Guide: Networking Essentials*, Indianapolis, IN, New Riders, 1997.

[10] D. Chafey, *Groupware, Workflow and Intranets: Reengineering the Enterprise with Collaborative Software,* 1998, Woburn, MA, Butterworth-Heinemann.

[11] K.M. Hussain, and D. Hussain, *Information Technology Management,* Bath, UK, Butterworth-Heinemann, 1997.

[12] P.G.W. Keen and M. Cummins, *Networks in Action,* Belmont, CA, Wadsworth Publishing, 1994.

[13] G.E. Bock and D.A. Marca, *Designing Groupware,* New York, NY, McGraw Hill, 1995.

[14] P. Lloyd, *Groupware in the 21st Century: Computer Supported Cooperative Working Toward the Millennium,* Westport, CT, Praeger, 1994.

[15] J.L. Creighton, and J.W.R. Adam, *Cyber Meeting: How to Link Peopl and Technology in your Organization*, New York, NY, AMACOM, 1998.

[16] R.D. Galliers, and W.R.J. Baets eds., *Information Technology and Organizational Transformation: Innovation for the 21st Century Organization, The Role of Information Technology in Organizational Transformation*, ed. J.A. Turner, John Wiley & Sons, Ltd., West Sussex, UK, 1998.

[17] J. Liebowitz, ed. *Knowledge Management Handbook, Groupware: Collaboration and Knowledge Sharing*, ed. G. Coleman, CRC Press, Boca Raton, FL, 1999.

[18] M.L. Markus and R.I. Benjamin, "The Magic Bullet Theory in IT-enabled Transformation," *Sloan Management Review*, 1997, 38(2), pp. 55-68.

Chapter 3

LOCATION-BASED SERVICES

Prabuddha Biswas, Michael Horhammer, Song Han, Charles Murray
Oracle Corporation

Abstract: This chapter describes a framework for providing location-based services based on the separation of content and content providers. Content from multiple providers can be aggregated and selected based on criteria reflecting what the customer has specified or purchased. Service proxies are provided to bridge the gap between the service provider's application and the providers of such content as maps, driving directions, and Yellow Pages listings. The framework provides a management infrastructure that allows the service providers to control which content provider is accessed. The framework also enables the organization and access of services by geographic regions. The chapter describes some demonstration applications that have been developed to demonstrate how interesting services can be developed using content from different providers. It also discusses several issues relating to location-based services that merit further work and study, and several areas that show promise for future development.

Keywords: Location-based services, Mobile applications, Region modeling

1.   SUMMARY

Some mobile services can be location-based with respect to their visibility, their content, or both. That is, a service might be visible only within certain specified geographical areas, and the content of a service might be tailored to those areas. For example, subway schedules and station information for the Bay Area Rapid Transit (BART) System might be available (visible) only in the San Francisco metropolitan area; and specific restaurants whose listings or advertisements appear on a mobile device (that is, the content) might depend on the user's location with respect to participating restaurants.

This chapter describes a framework for providing such location-based services. This framework is based on the separation of content and content providers, whereby content from multiple providers can be aggregated and selected based on criteria reflecting what the customer has specified or purchased. Service proxies are provided to bridge the gap between the application and the providers of such content as maps, driving directions, and business directory listings. The service proxies use a flexible multiplexer engine to provide fault tolerance, customizable failover mechanisms, and reliability control among various providers.

This chapter describes some demonstration applications that have been developed to demonstrate the usefulness of our framework. One is a business-to-consumer application for recalling information associated by location-given incomplete data and some measure of proximity, a list of locations can be retrieved (for example, a winery and a sushi restaurant within one kilometer of each other in a metropolitan area). Another is an enterprise application for delivery routing and dynamic rerouting-given pickup and delivery orders and a fleet of trucks, determine the optimal route for each driver at the start of the day, and adjust routes during the day as important pickup orders are received.

This chapter also discusses several issues relating to location-based services that merit further work and study, and several areas that show promise for future development.

2.   INTRODUCTION

The explosive growth of the use of intelligent mobile devices such as mobile phones and Personal Digital Assistant (PDAs) directly opens up a whole new world of possibilities: delivering location information to those mobile devices, and delivering services to the mobile devices that are customized and tailored according to their current location. In this chapter we define the problem, discuss the existing technologies, and present a framework that adds power and flexibility.

3.   PROBLEM DEFINITION

We are observing a convergence of PDAs and mobile phones, such as the Ericsson R380 [1] and Kyocera QCP 6035 [2]. PDAs are acquiring advanced wireless communication capabilities, and mobile phones are getting more computation and Internet access features. In the near future most appliances will also have network interfaces and remote management capabilities [3]. These advances, together with inexpensive position-sensing devices, will make location-aware applications popular. However, due to the lack of an appropriate location-aware application development framework, many basic services will be reimplemented in each application, and cooperation between different location-aware applications will be difficult.

Today there are also a number of vendors that provide specialized location services such as geocoding, mobile positioning, mapping, routing and business directories. Each of them has its their own advantages and shortcomings. For example, mapping provider A may provide maps of better quality in the US but lacks coverage in Europe. Routing provider B's service may have a faster response but be more costly than provider C's. The motivation behind our work is to provide a framework that enables building complex location-based applications from the building blocks of different providers and sources. It is more useful if the framework can provide a location-dependent service such as a "Local city guide" that automatically provides contents depending on the user's location.

Prototype implementations of location-based services have been attempted on a small scale like a building or city [4,5,6], but commercial applications today in general do not take into account the location information of the clients or the service providers. By providing an infrastructure that captures the mapping of services in areas of interest (or, regions) and a mapping of users to the regions, the users can be presented with information relevant to their location. Applications can determine the static locations of mobile business objects (services or clients) as well as the path taken by the mobile objects in a specified time window; and with this type of information, developers can create innovative location-aware applications.

4. OPPORTUNITIES

Many mobile device users are attracted by new, specialized features, and they are willing to pay premium prices for the right services. Auctions of cellular bandwidth have proven that new devices and services are expected to play a major role, and that no service provider can afford to ignore this trend. Providers are anxious to participate in "the next big wave," even if the exact expected profit is currently difficult to estimate.

Given a technology environment marked by constant and unpredictable change, as well as a business environment with increasing capital at risk, mobile service providers are anxious to protect their investments. They ask such questions as: Can we be prepared for the next technology beyond streaming video on a color cell phone? Can we accurately estimate the worth of bandwidth to be made available at the next auction? And can we limit our exposure to loss of service from an external provider that we currently depend on?

5.   FRAMEWORK FOR LOCATION-BASED APPLICATIONS

The location-based application framework we propose is based on an application server that serves any content to any device. Applications are independent of the target device, yet automatically can exploit specific features of the device and provide customized content depending on the availability of such features. The benefits of this approach include:
–   Better protection of investment. It is redundant and expensive to develop separate platforms and applications for each new device.
–   Faster conformance to the newest device technology. The first players in a new device technology tend to have a competitive advantage, and this flexible framework makes it easier to quickly adapt to new device technologies.
–   Faster acceptance of new devices due to instant support of existing services. Fast acceptance reduces bound capital and the length of amortization.

Figure 1 illustrates the proposed base application server framework, including its inputs and outputs.

As shown in Figure 1, the core of the framework is a module that performs a variety of services: adapting, general processing, and transforming. This module can obtain input from any source, adapt it, and represent the data in an intermediate XML format. It also handles general tasks, such as authentication and messaging. To interface with the various devices, it transforms (or, translates) the intermediate XML into the markup language (a specialized ML) of the target device. Although the use of XML for representation and transformation incurs a minor processing overhead, XML is essential to ensure the flexibility of the framework and its extensibility to future device technology.

5.1   Any Content

Applications using the base framework can use any content that is available from the Web, a database, or the file system. Applications can be programmed in any language. Opportunities for reusing code and content are considerable, and the advantages for development time and cost are apparent.

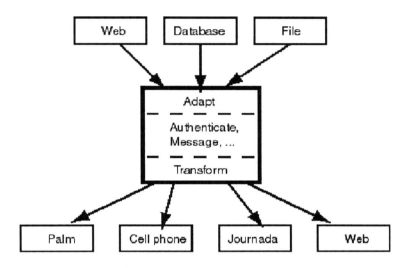

Figure 1. Application server framework

5.2 Any Device

The framework uses an XML intermediate format for the exchange of data between the data providers and the application, and it uses specialized ML (such as VoiceML) for output to target devices. XML is designed to support any type of content of any complexity, and to support presentation on a variety of devices. Obviously, devices vary in their ability to display certain content in a reasonable manner, or even to store it in their memory. For example, an application that presents high-resolution streaming video can be used only on a sophisticated device, and voice recognition cannot be used on a Palm device without a microphone. However, by designing the output XML carefully, application developers can enable different input and output options depending on a specific device's capability.

The framework will by default exploit the maximum hardware capability of the device to present information. If the application does not care about hardware capabilities, the device will do its best to present the information. Alternatively, an application can determine the device type and capabilities and can optimize the content accordingly. For example, a movie-review application might have both a short and long form of the review for each movie, and choose which form to display depending on the device characteristics. If the movie-review application chose to ignore device characteristics and displayed one (long) form in all cases, a "smart" cell phone would then automatically distribute the content over several screens, and perhaps successively download only parts of it to reduce memory consumption.

Most importantly, applications that work on today's devices will continue to work without limitation with tomorrow's more advanced devices and markup languages.

5.3   Anywhere

Mobile devices naturally benefit from spatial awareness, especially when supported by mobile positioning technology. An application can choose whether to base location on automatic positioning or manual positioning. Automatic positioning uses information from the mobile network or from the GPS in the device to detect the user's physical location. The framework allows each user to choose certain locations (such as home, office, and local airport), and to designate any of them as the default - this is referred to as manual positioning. Alternatively, it might let users enter an arbitrary location. Manual positioning can be used when the automatic positioning is not available or not relevant. Manual positioning is useful if users want to have the "what-if" option, for example, to be able to view all the services they will see in the San Francisco area even though they are physically still in Boston.

5.4   Any Provider

Spatial support of wireless devices is incomplete without specialized services, such as geocoding, reverse geocoding, driving directions (routing), yellow pages, white pages, maps, weather forecasts, traffic reports, and demographic information. The technology for all these services is available, and in some cases even mature. However, problems arise if an application provider is forced to commit to one of the available data providers: lack of flexibility, high service fees for "locked-in" customers, and exposure to denial of service for technical, business, or other reasons.

The location-based services framework provides independence from any specific data or service provider. An application can use data from an internal (in-house) provider, if there is one. Alternatively, it can use data from one or more external providers. Generally, though, a provider of a location-based service does not possess the resources, technology, or data to provide all services in-house. The framework, then, serves as a broker to a variety of external providers and lets application providers choose and quickly shift among external providers, thereby making the decision to rely on external providers easier and more palatable. The result should be mutually beneficial to both parties: a "buyer's market" aiding application developers, and the possibility of a larger market benefiting the data and service providers.

6. LOCATION-BASED SERVICES

The location-based services framework supports access to a number of spatial services, such as geocoding, driving directions, yellow pages, and mapping. Typically, the framework does not perform the services itself, but instead stores information about providers and performs multiplexing between them and the application. Provider information is stored in XML format, as illustrated in the following example:

```
<?xml version="1.0 standalone="yes">
<Providers>
  <Provider
    ProviderName = "ProviderA"
    ProviderImpl = "oracle.panama.spatial.router.ProviderARouterImpl"
    URL       = "www.providerA.com"
    UserName   = "..."
    UserPassword = "..."
    Parameters  = ""/>
  <Provider
    ProviderName = "ProviderB"
    ProviderImpl = "oracle.panama.spatial.router.ProviderBRouterImpl"
    URL       = "www.providerB.com"
    UserName   = "..."
    UserPassword = "..."
    Parameters  = ""/>
  <Provider
    ProviderName = "ProviderC"
    ProviderImpl = "oracle.panama.spatial.router.ProviderCRouterImpl"
    URL       = "www.providerC.com"
    UserName   = "..."
    UserPassword = "..."
    Parameters  = ""/>
  ...
</Providers>
```

The multiplexing approach provides several technical and business-related advantages:

– Integration of internal and external services: Any existing in-house solution (such as a yellow page database) can be easily integrated with suitable external services. For example, the framework might be configured so that an external yellow page provider be accessed only when the internal yellow pages database is unavailable or overloaded, or does not have the data to satisfy the request.

– Business flexibility and risk abatement: Relying solely on one external provider for a service amounts to a commitment that can become expensive and risky.

– Code simplicity: Without the multiplexing framework, more code (and more complex code) would be required to handle a portfolio of several external provider directly.

– Cross-application communication: Information exchange between services, such as between yellow pages and driving directions, is simplified because of the common framework. Figure 2 shows the information exchange among geocoding, mapping, routine, and yellow pages services in an example application.

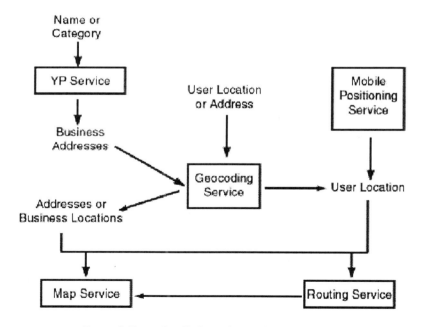

Figure 2. Example of information exchange among services

The following sections describe some of the services in more detail, including problematic issues associated with each.

6.1 Geocoding

Geocoding determines the longitude and latitude coordinates of an address. Geocoding is the most fundamental location service, because it is used directly or indirectly by all the other services. The technical requirements of a geocoding service can range from trivial to demanding. An approach is to disregard any address fields except the postal code and return the coordinates of the center point (centroid) of the postal code area. This can be done efficiently by a simple database query supported by a common B-tree index. An advanced service might assign different coordinates to every house number in the same street, and might reliably determine which side of the street the house is on, and whether a given number even exists on the street.

Most existing geocoders can provide more detailed information than just a postal code centroid, but typically not the most demanding information. In many cases, any claim regarding the side of the street must be taken with care. For example, providers of driving directions sometimes use the side of the street to determine whether the user should turn left or right out of the driveway. If the geocoder cannot reliably determine the side of the street for a location, the initial turn in the routine instructions may send the driver in the wrong direction.

6.2 Mapping

Mapping enables users who have devices with graphical capabilities to visualize location-related data. Technically, mapping usually involves spatial database queries and advanced visualization algorithms. These operations are computationally demanding and depend on large geographic databases. Existing providers differ in the scalability of their approach, the level of map detail, and the visual appeal of their maps. Conceptually, however, mapping is well understood.

6.3 Routing

Routing is more commonly known as *driving directions*. Technically, routing is equivalent to graph searching, which is actively being pursued in artificial intelligence and other fields. Given an effective, efficient and optimistic heuristic of driving distance, A* is a good example of a routing algorithm.

In addition to turn-by-turn instructions, routers might also provide maps of each turn and of the complete route. The router might even supply a list of point coordinates along the route, to enable the requesting user to perform

some spatial analysis. For example, the user can identify the customers
he/she can visit along the route. Routing maps generally offer fewer options
than general maps: for example, users typically cannot adjust the map scale.
Routing maps are automatically centered around a route or a single
maneuver (turn).

6.4   Yellow Pages

Yellow pages is a service that can determine a list of businesses matching
a specified region and either a business name or a category. The technical
aspects of a yellow page implementation are well understood: it involves
common database queries, and the speed and complexity of the solution
depend on the level of database optimization. However, the semantic aspects
of categorization pose more challenges.

Semantic categorization of yellow page data has not yet attained maturity
or standardization. Different providers use different approaches. Of course,
they use different business categories with differing semantics and names,
each in an effort to add value and establish their own brand recognition.
More importantly, the organization of categories in hierarchies varies
widely, between flat lists and deep hierarchies, balanced and unbalanced
trees, fan-out ratios of 5 to 100. Even non-tree graphs can be used. Keyword
searches of categories might be semantic-based, substring-based, or
nonexistent. For example, a semantic-based search for *restaurant* would
return all categories containing *restaurant* (such as *restaurant equipment*)
but also *Chinese cuisine*, whereas a substring-based search would return only
categories containing the search term.

The location-based services framework attempts to solve the semantic
categorization problems by letting administrators create local custom
hierarchies based on their own semantic preferences. This custom hierarchy
is defined in an XML document and is completely independent of the
existing categories of any external provider. Each category in the custom
hierarchy is mapped to one or more external provider categories that have at
least similar semantics. This approach unifies the category hierarchies of
external providers and gives the application implementer the flexibility to
use its own branding of categories.

The following example shows a yellow page custom hierarchy definition
in XML format.

```
<?xml version="1.0 standalone="yes">
<Categories>
  <Category CategoryName = "Restaurant">
    <Provider
      Name = ";ProviderA;"
```

```
        Parameter = "..."/>
      <Provider
        Name = ";LocalDB;ProviderB;"
        Parameter = "12345"/>
      </Category>
      <Category CategoryName = "Hotel">
       <Provider
        Name = ";LocalDB;ProviderB;"
        Parameter = "45678"/>
      </Category>
       ...
      </Categories>
```

7. AUTOMATICALLY LOCATING MOBILE USER

The positioning information is critical to many location-based application scenarios, such as:

- Identifying relevant location-dependent services. Mobile positioning is required to automatically determine which location-dependent services are important to the user at his or her current location.
- Dynamic routing-routing from a user's current location to a destination.
- Business query-looking for businesses that are close to a user's current location.

A mobile user can manually position himself or herself by specifying a street address, a postal code, or other geographical information. A user's position can also be provided by the mobile network. The performance and scalability of network-based mobile positioning has been researched in great detail [7,8,9,10].

Today a number of vendors can provide mobile positioning services over the Internet [1,11]. Ideally these services should have a standardized access interface, but one is not currently available. The mobile positioning component provided by our framework enables the developers to access a mobile target's real-time positioning through a unified programming interface. Further, there is an in-memory cache for better performance and scalability. The location information can also be archived on disk in a location log for future analysis.

People are also concerned about the privacy of the location information. Our framework will provide a hierarchical, end-user customizable privacy control for mobile positioning. Users can choose not to be positioned, or to be positioned by individual users or groups. For each level of positioning privilege, the user can specify a time range. Further, users can turn off location cache and location log features if they do not want location information to be stored in memory or persistent storage.

8.   LOCATING AND ORGANIZING SERVICES

Providing services to mobile devices is more than just converting Web information to match the characteristics and limitations of the device, such as the small screen and low bandwidth. A typical Web portal, as provided by companies such as Yahoo! or as provided internally by corporations to their workforce, is loaded with as much information as possible, usually in the form of jumping-off points to other information repositories. This approach is too limited when one considers mobile devices that, by definition, can be anywhere. However, when one adds the capability of the Web server to recognize the geographical location of a mobile device, the server can then adapt itself and provide services or information that are relevant to the current location.

Extending this model, services can then be built with content that is more or less local to specific locations. For example, one service might monitor local buses and trains for a specified city, while another might monitor inter-city trains and commuter airline traffic. This model can extend to local restaurants advertising their menu, hotels announcing room availability, theaters listing last-minute tickets for sale, and many other applications. Eventually, services could themselves be physical devices that just advertise their availability, such as public printers on a campus, ATMs, and vending machines. Clearly, in such a model the number of services potentially relevant for a given location can grow enormously and be very dynamic, with new services appearing and disappearing constantly.

Figure 3 illustrates such a model, in which the location of the user and the location of the service determine which services are relevant to a user at a given location.

As shown in Figure 3, the location of the user can be determined by or interpreted as any of the following, depending on the application:
- The user's actual current position, determined dynamically by the mobile network
- A location mark (for example, *Home* or *Office*) defined by the user
- Any location specified by the user (not necessarily the current position)
  Moreover, the location of the service can be determined by or interpreted as any of the following, depending on the application:
- An administrative boundary defined by the service provider (for example, *Greater Chicago area* or *All Massachusetts state parks*)

– A specific address or region around an address (for example, *Atlanta Airport* or *5 miles around my office*). A custom region can also be defined, such as the *New England states* by aggregating the relevant states.

The next section describes an implementation of this model.

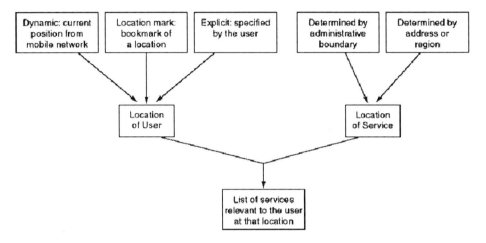

Figure 3. Location of user and service

9. IMPLEMENTATION USING REGION MODELING

Various efforts have been made to provide location-dependent frameworks and applications. Some frameworks enable location dependent applications in a relatively small range such as a room, a building, or a campus [4,5]. People are also prototyping simple location based applications in the range of a city [5,6]. In these works, they use simple simulation of a wireless network and provide interesting location-related applications such as city tourist guide information. When determining whether a service or a tourism attraction should be available to a user, the only factor that is considered is address or distance from a landmark, instead of more powerful, hierarchically organized geographical regions. However, when you try to deliver a commercial application of this type with a wider coverage, you need to be able to efficiently organize and model world-wide regions in databases, integrate different mobile positioning technologies, get real business information from various business directories providers, and get maps from providers that have international coverage.

Our proposed implementation adds flexibility to the concept of a location-aware service, by allowing for location relevance or availability to be an attribute of the service and by allowing automatic detection as well as manual specification of the user's location.

9.1   Services and Regions

Our current implementation associates services with geographical areas, or regions. A region can of course be defined as a member of a hierarchical structure (country, state or province, county, city or municipality, district, street, and so on). However, this model is too simple to represent reality, where regions are likely to be defined in more flexible ways, crossing over the hierarchical boundaries. For example, a metropolitan area can spread over state or even country boundaries, or a company-specific sales area can cover areas over multiple counties, or the path of a bus can follow multiple roads irrespective of their location.

We propose a region modeling scheme that provides system-defined regions and allows user-defined regions. In the system-defined region model, the entire world is divided into sub-regions with the hierarchy shown in Figure 4.

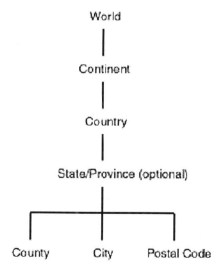

Figure 4. Hierarchy in system-defined regions

In addition to the system-defined regions, one could create user-defined regions that can contain and/or intersect multiple system-defined regions.

Given a particular set of longitude/latitude coordinates, we can easily determine all the relevant regions at different levels of the hierarchy. Also, given a particular user-defined region, we can find all system-defined regions that are totally contained in it at a particular level of hierarchy, as well as across different levels of hierarchy. We will also store the metadata associated with the different business objects and their locations (location and metadata mapping). Therefore, given the location information and metadata criteria, we can get all the different businesses in that region satisfying the metadata category.

The region hierarchy and the metadata information will be stored in the database. These will form the basis for the location-aware application framework. In such a model, it is essential for mobile users to be provided with the means to filter and prioritize the relevant services.

9.2 Service Discovery

Our proposal goes beyond the limitations of many current services by allowing location to be an inherent attribute of a service and by allowing automatic as well as manual discovery of relevant services.

Many existing services that are called "location-aware" accept one or more location-related input parameters. For example, a weather service might ask the user to input a postal code. The service then checks an internal table that associates each postal code with the nearest weather station, and it presents weather for that station.

With our planned implementation, location relevance or availability is added as an attribute of the service. The location attribute is in addition to a service's ability to accept optional input parameters. For example, a BART (Bay Area Rapid Transit) train schedule service would have as its location attribute San Francisco and the surrounding area. That is, BART subway schedules are relevant only within that location. In addition, the service could allow several optional input parameters, such as a station name for users who want to know when the next train to a particular station leaves from their current station.

Table 1 provides additional examples of services that could be created using our implementation.

The location, moreover, could be automatically detected or manually specified, or the service could allow either option.

- With automatic detection, if the user's position is within the service's region, the service would be automatically enabled. For example, if the user had subscribed to a public transportation service, BART schedules would be available when he or she enters the Bay Area region.

− With manually specification, the location of interest could be a user-defined location mark (for example, *My house*), or it could be some other location specified in any of several ways (such as a city, street intersection, street address, or set of longitude/latitude coordinates).

Table 1. Services with location attributes: Additional examples

Service	Location Attribute	Optional Parameters
Virginia attractions	Virginia (U.S. state)	Type (history, arts, specialized categories
U.S. sport stadiums and arenas	Selected points across the United States (user-defined)	sport name; team name
Art museums	Selected cities (user-defined)	genre; event type

9.3   Associating a Region with a Service

When you define a service, you can also specify a location attribute, that is, associate a region with the service. For example, in the following prototype implementation shown in Figure 5, to associate a region with the service being created, check (enable) the Location Dependent box.

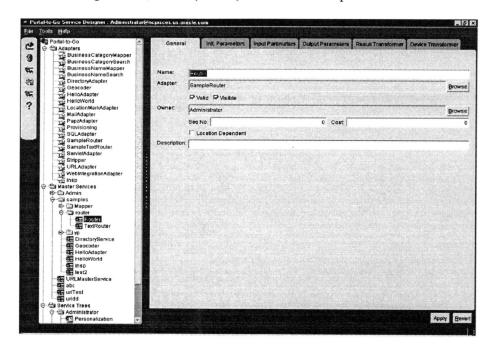

Figure 5. Creating a Service

When you specify that a service is location-dependent, you can use a region modeling tool to select a region to be associated with the service, as described in the next section.

9.4   Region Modeling Tool

A region modeling tool lets administrators of a wireless portal service identify specific areas for services, as shown in the illustration of a prototype in Figure 6.

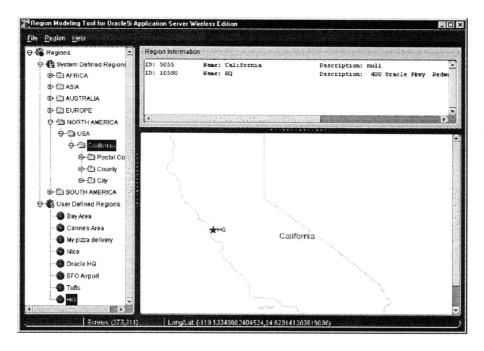

Figure 6. Region modeling tool: Associating a region with a service

In Figure 5, the user has selected a previously created user-defined region named *HQ* (for Oracle headquarters) to be associated with the service.

A region is simply a geographic entity, or location. A region can be small (such as a street address) or large (such as a country or continent). A region can be represented by a point, as is often done for addresses and locations of interest (such as airports and museums), or by a polygon, as is usually done for states and countries.

Service providers may want to define specific regions for a variety of applications and services, such as:

- City guides for selected metropolitan areas, so that users in those areas receive only services and information (such as restaurant listings or advertisements) relevant to them.
- Art museums in a city or a multistate area, so that art lovers can plan trips to museums.

A company may provide many specialized services, and customers may be able to subscribe to and pay for individual services tied to regions. For example, one customer might subscribe to city guides for the entire United States, while another customer might subscribe only to city guides for southeastern states.

9.4.1 Folders and Hierarchies of Regions

Regions are stored in folders. Folders can be organized in a hierarchical fashion. There are two top-level folders: System-Defined Regions and User-Defined Regions:

- System-defined regions are arranged in a hierarchy of predefined areas: continents, which contain countries. The United States contains states, which contain postal codes, counties, and cities.
- User-defined regions are regions created by users, based on entering an address or on selecting one or more other regions (system-defined or user-defined).

10. DEMONSTRATION APPLICATIONS

Rough prototypes have been developed of two applications that use the location-based services framework: a business-to-consumer application called "Location Recall," and an enterprise application providing static and dynamic routing of delivery trucks. These applications are designed to provide proof of concept. By demonstrating practical uses of the framework on a limited scale, they will show the feasibility of the technology and its probable value on larger scales.

10.1 Location Recall

The location recall application finds businesses based on names or categories and on spatial associations. For example, a user might remember visiting a Gallo winery near Los Angeles and later dining at a Sushi restaurant not far from the winery. (Or, a user might want to visit such a winery and then dine at a Sushi restaurant.) The user could ask the application for all pairs of a Gallo winery and a Sushi restaurant within a specified distance of each other in the Los Angeles area. The application performs the following logic:

1. Within the Los Angeles area, perform a yellow page lookup by name for *Gallo*, and store all winery locations.
2. Within the Los Angeles area, perform a yellow page lookup by category for *Sushi restaurants*, and store all matches.
3. Geocode all locations found in the above two queries.
4. Perform spatial analysis to find all pairings where a winery and a restaurant are within the specified distance of each other.
5. Generate driving directions with maps: from the user's current or specified location, to a chosen business in a chosen cluster.

This application uses the YP, mapping, routing and geocoding services, but appears to the consumer as a single application.

10.2 Delivery Routing

The delivery routing application assumes a package delivery company with pickups and deliveries and a fleet of trucks. At the start of each day or shift, trucks are loaded and the drivers are sent out with a list of scheduled pickups and deliveries. At varying times during the day, important pickups are added, and drivers can receive amended route instructions when they check in after each stop. The application performs the following logic:

1. Before the start of the day or shift, geocode the addresses of all scheduled pickups and deliveries.
2. Perform spatial analysis to allocate the pickups and deliveries among the available trucks in a sensible manner using cluster analysis
3. Generate an optimal route with driving directions and maps for each driver.
4. As new pickup orders are received by phone or the Internet, find the most appropriate driver in each case and modify the remainder of the route (after the next scheduled stop).
5. Generate a revised optimal route with driving directions and maps for drivers affected by the changes. (Drivers check their mobile devices for any messages or changes after each stop.)

The application could be enhanced in the future to add greater flexibility and value to the business. For example, it could add support for traffic services to allow rerouting for scheduled activities such as construction and unscheduled events such as accidents and fires. It could add support for weather services to adjust routing and schedules for such things as snowstorms and flooding. It could allow for instant alerts and rerouting in critical situations, as opposed to waiting until the next stop.

11.   ISSUES AND FUTURE DIRECTIONS

Several issues relating to location-based services merit further work and study, and several areas show promise for future development. This section explores some of these issues and areas.

11.1   Location-Based Push Services

When you arrive in a foreign city, you may be approached by a "middleman" who wants to take you to a hotel. This person gets paid by the hotel if you agree to stay in that hotel. The same principle could be used where businesses would be willing to pay for pushing targeted messages to clients looking for a specific service. For example, if you perform a query looking for Italian restaurants, Italian restaurants in the neighborhood could send you a "mobile coupons" offering special deals.

11.2   Location-Based Events

If you were a busy executive and needed to meet with two of your colleagues as soon as they arrived at the office, you could define a "region" around your office and define an "event" that triggers when both colleagues are detected to be within the defined office region. The action of the event could be to send messages to the colleagues requesting them to stop by your office for a meeting.

11.3   Geocoding Within Small Distances

Geocoding services currently provide varying levels or detail and precision. This affects an application's ability to do spatial analysis involving small distances. For example, driving directions might not accurately indicate whether to turn left or right at the start or end of a route. This is due to the difficulty of determining which side of the street a geocoded address is on. Geocoders often claim to know the side of the street, but cannot always (correctly) deliver on that promise.

Furthermore, a mobile device might determine that it is close to a particular landmark or business, but it might not reliably state the direction.

It would be interesting to improve driving directions with hints, such as "Turn left onto Main Street (10 meters after passing a Shell gas station on the right-hand side)." This would require the service to know about the gas station and whether it is on the left or right side of the street. Furthermore, reliable knowledge is required of whether the gas station is directly before or directly after the proposed turn.

We assume that geocoding services will eventually support the level of detail needed to provide such features.

11.4   Provider Selection Rule Engine

As a possible improvement to the provider preference list, a simple rule engine could support more flexibility. Rules could define a provider's fitness based on region, language preference, time, and possibly other parameters. Provider A might work well in the US, while it might not support Europe. If the rules are kept simple (for example, restricted to horn clauses), the rules engine can very efficiently deduce a dynamic provider ranking. A very general rules framework might adversely affect the performance of the system.

11.5   Mapping and Image Transformation

Several mapping providers deliver impressive results, both in the level of map detail and the visual appeal. However, most provide map images only in GIF format, and they do not support BMP (bit-mapped) and WBMP (wireless bit-mapped) format, which are especially important for mobile devices (many devices support only one of these formats). What is needed is support for full image transformation, not merely conversion from GIF to BMP/WBMP and back, but other operations such as scaling, flipping, and rotating of images.

The location-based services framework provides such image transformation; however, ideally this feature should be performed by the mapping provider itself. The mapping provider has to create an image anyway and serve it to the client; and if the client needs BMP or WBMP format, the mapping service incurs no significant additional overhead or complexity in providing the desired format as opposed to GIF. Moreover, if the mapping service provides BMP or WBMP format directly to the device, the image data is transferred over the Web just once; whereas if a GIF image is provided to the framework for transformation and then provided to the device, the image data is transferred over the Web twice -- thus doubling the network traffic related to the image.

There is no benefit in first representing all image content in a particular format (e.g. GIF) and then transforming it to the target format. It would be good if mapping and routing providers support multiple image formats, including BMP and WBMP, on their servers.

11.6   Location-Based Content: What's New?

11.6.1   Demographics

Demographics could be an important addition to the location-based services framework. Applications will then be able to access a wide range of statistical data for any given location. The granularity of the data will probably be by postal code, making the service technologically straightforward for any external provider. Demographics would be very useful for targeted advertisements, real estate applications, business site finders, and other applications.

11.6.2   Traffic and Parking

Traffic and parking information would also be valuable for location-based services. This information can directly benefit the quality of driving directions. However, traffic information is more technically demanding than demographics for several reasons. Granularity at the street segment level is much smaller than at the postal code level. Also, referencing a street segment with a unique ID is more complex than for a postal code, where the name itself serves as an ID. Furthermore, traffic and parking data must be updated more frequently. Finally, for driving directions to benefit from traffic information, a detailed list of street segments has to be queried and compared with the complete route.

11.6.3   Weather Information

Support for weather information is also interesting, and this should be easier to integrate into the framework than traffic information. Weather information is usually less dynamic than traffic information, and the granularity is likely to be much coarser (probably at the postal code or city level). For example, it will probably be sufficient to know that it will rain heavily in Boston this afternoon, but Boston traffic information will need to be updated more frequently and identify specific locations for problems such as accidents and breakdowns. One could also look at a driving direction and predict weather along the route. For example, for a person traveling from New York City to Buffalo, New York in the winter, it might be useful to input the departure time and find out that there will be a blizzard in Buffalo when that person reaches there (no surprise!).

11.7 Factors Affecting Complexity

Location-based services, both current offerings and any future enhancements, vary considerably in their algorithmic complexity. A simple query such as finding businesses in a postal code places minimal algorithmic demands. However, the demands become much greater if multiple dimensions are introduced or if complex spatial analysis is involved, or both. Multiple dimensions might involve such things as requesting businesses not only by yellow pages category but in a certain size or revenue category as well. Complex spatial analysis might involve limiting the results to cases where locations are within a certain distance of one or more points.

The extent to which data is static or dynamic varies among the kinds of data involved. Geocoding and mapping data is perhaps most static, since addresses and landmarks tend to be constant. Routing information is fairly static, although construction affects the status and the number of roads available. Demographic and yellow pages information is somewhat static, though the rate of any population and business changes affects its accuracy and value. Weather information is relatively dynamic, and traffic information is highly dynamic.

12. CONCLUSION

We have presented a framework to enable the development of location-based applications. The principal recurring themes are extensibility and flexibility with respect to future mobile device technologies. An application can easily integrate services from internal and external providers. The application is not dependent on specific providers because the framework acts as a broker or multiplexer. Because multiple providers can be accommodated, any failure by providers is transparent to the application and its users. The adoption of our proposed framework will, we believe, bring substantial, tangible benefits to both providers and consumers of mobile services.

REFERENCES

[1] Ericsson Mobile Positioning System, http://www.ericsson.com/
[2] Kyocera Smartphone, http://www.kyocera-wireless.com
[3] Jini [TM], http://www.sun.com/jini/
[4] N. B. Priyantha, A. Chakraborty and H. Balakrishnan, "The Cricket Location-Support System," *Proc. MOBICOM 2000*, pp. 32-43.

[5]   V. Krishnan, "Location Awareness in HP's CoolTown," W3C Workshop on Position Dependent Services 2000.

[6]   K. Cheverst, N. Davies, K. Mitchell and A. Friday, "Experiences of Developing and Deploying a Context-Aware Tourist Guide: The GUIDE Project," *Proc. MOBICOM 2000*, pp. 20-31.

[7]   Z. Naor, "Tracking Mobile Users with Uncertain Parameters," *Proc. MOBICOM 2000*, pp. 110-119.

[8]   H. Levy and Z. Naor, "Active tracking: Locating Mobile Users in Personal Communication Service Networks," *Wireless Networks*, 1999, Vol. 5, No. 6, pp. 467-477.

[9]   U. Madhow, L. Honig and K. Steiglitz, "Optimization of Wireless Resources for Personal Communications Mobility Tracking," *IEEE Trans. on networking*, 1995, Vol. 3, No. 6, pp. 698-707.

[10] Z. Naor and H. Levy, "Minimize the Wireless Cost of Tracking Mobile Users: An Adaptive Threshold Scheme," *IEEE INFOCOM'98*, pp. 720-727.

[11] Nokia mPosition Platform, http://www.nokia.com

Chapter 4

AN ADAPTABLE NODE ARCHITECTURE FOR FUTURE WIRELESS NETWORKS

Donal O' Mahony, Linda E. Doyle
Trinity College, Dublin

Abstract: We describe an adaptable node architecture for future 4th Generation wireless networks. The node will be capable of running a large range of network applications and will be able to adapt its mode of operation to the radio environment and prevailing wireless network architecture. The core of this node consists of a dynamic protocol stack that can be configured, from a suite of protocols, based on the requirements of the underlying network. The node will be capable of operating in traditional wireless networks as well as in Ad-hoc wireless communications systems.

Keywords: Adaptable architecture, 4th Generation communications, Ad-hoc networks

1. INTRODUCTION

1.1 An Adaptable Architecture for 4th Generation Systems

The first generation of mobile telephones were based on analogue technology and were relatively unsophisticated. They were followed by 2nd generation systems which involved digital technology. The most successful of these 2nd generation systems is undoubtedly the European Global System for Mobile (GSM) standard. Work is now ongoing in the area of 3rd generation mobile systems. In Europe, 3rd generation systems are referred to as the Universal Mobile Telecommunications Systems (UMTS) whereas the global standardization effort undertaken by the ITU is called IMT-2000. The direction of this work is to continue to evolve today's circuit switched core network to support new spectrum allocations and higher bandwidth capability and to place greater emphasis on the provision of data access [1]. Efforts are also being made to integrate the many diverse mobile environments in addition to blurring the distinction between the fixed and mobile networks.

At Trinity College, Dublin, we are pursuing a line of research that goes beyond 3rd Generation mobile telephony into what we call 4th Generation Mobile Systems. This research effort aims to define a replacement for the global fixed and mobile telephone networks with a core based on Internet Protocol (IP) technology and supporting many different kinds of wireless access networks. This chapter describes an adaptable node architecture for these future wireless networks.

Technological advances have made possible the production of low-cost computing devices equipped with wireless digital communications. Considerable research is taking place at present, both in the academic and commercial research communities and also in the marketplace to determine the form of the wireless networks of the future. So many questions are still open. What radio spectrum will be best? What kind of modulation is most appropriate? How will access to the spectrum be regulated? What will be the overall architecture of the system – cellular or Ad-hoc? How will nodes be identified? How will information be routed to its destination? What kinds of security measures need to be taken?

It is clear that there will be multiple answers to these questions depending on the intended application. What is appropriate for communication between a swarm of sensors collaborating to process information may be quite different to that used in a future mobile phone replacement system. We believe that the key to designing future wireless systems is in the creation of an adaptable wireless node.

We have been focussing on a software and hardware architecture that will allow us to experiment with many different facets of wireless networking from mobile applications through to core radio issues using a component-based approach. The ultimate goal is to produce a general-purpose mobile node capable of running a large range of network applications that can adapt its mode of operation to the radio environment and prevailing wireless network architecture.

1.2   Overview of Chapter

Section 2 outlines the core architecture of the wireless node. Section 3 details the desired characteristics of a 4th Generation Mobile Node and describes how that node is realized using the core architecture. Section 4 gives details of the security architecture for 4th Generation wireless nodes and Section 5 summarizes the chapter.

2.   THE LAYER CONSTRUCT

2.1   Layered Reference Models

The use of layered reference models as a means of decomposing complex networked systems has been with us since IBM's Systems Network Architecture (SNA) and the ISO Reference Model for Open Systems Interconnection (OSI) [2]. The layer boundary allowed an abstraction barrier to be constructed around the functionality of the layer and, in the case of OSI, the precise primitives used for inter-layer communication and their associated parameters could be precisely specified. The use of abstract primitives in OSI allowed implementers a choice of what real programming constructs (e.g. subroutines, tasks, processes) to use to implement the layers themselves and also what mechanisms (e.g. sub-routine calls, inter-process messages) are used to communicate between them. The final choice was generally determined by the native facilities available in the operating systems and by the language chosen for implementation. In logical terms, layers maintained their own state and implicitly executed their own state machines in parallel with all other layers that were present. One of the common problems for implementers was to map this implicit assumption into an environment supporting a single thread of control.

In the UNIX streams architecture [3], a very simple inter-layer interface was formulated that basically involved passing a 'message-block' between layers. Most of the time, this block contained data to be sent on the link but it could also contain control directives and parameters. Clark [4] argued that the overhead associated with data copying of this kind of structure may be very large in relation to that used in implementing the control operations of a particular protocol.

In order to avail of the large number of commercially written applications, we wanted to choose a commodity computing architecture. Since we were interested in communicating between palm-top and hand-held devices as well as fixed workstations, the logical software environment to use was Microsoft Windows. Different variants of this base operating system ran on devices with all popular form-factors, and a subset of the Win32 application programming interface (API) was available across all platforms.

The Win32 operating environment has a native implementation of threads available on all relevant platforms and this allows the implicit multitasking inherent in logical layered architectures to be naturally represented in code. Thus each layer is represented by (at least) one thread and since the operating system will pre-emptively schedule each thread to

receive a fair share of processor resources, the networking programmer need
not worry about scheduling issues.

2.2   Implementation

We have adopted a highly simplified version of the UNIX streams
method of inter-layer communication. Figure 1 shows a block diagram of
our generic layer structure.

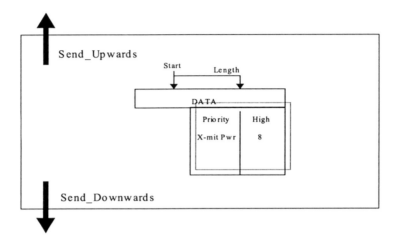

Figure 1. Generic layer structure

Our message blocks are dynamically allocated at the source (usually in
the application or at the network interface) and a reference to them is passed
from layer-to-layer without the need for copying. Data primitives tend to
grow as they progress downwards through a protocol stack as they acquire
header and trailer fields and undergo corresponding shrinkage as they rise up
a stack. Our message blocks are sized to deal with the maximum sized data
unit with pointers to indicate the beginning and size of the currently active
area of the buffer. This allows individual layers to add and strip-off headers
and trailers without requiring data movement. Having a single size for the
message block simplifies memory allocation and disposal. When a layer has
processed a message block, it invokes one of two primitives: *send_upwards*
or *send_downwards*, which pass the block to a neighbouring layer. A
message block is automatically de-allocated when it 'falls off the end' of a
stack.

Data giving information about the circumstances of the data reception or
directives on how it should be processed are contained in a 'blackboard' type

structure attached to each message block. This is structured as an arbitrary set of attribute value pairs and can be added-to, modified or interrogated by any layer. In this way, it is possible for an application to specify information that may be acted upon by a layer much lower down the stack providing extra facilities for Application Level Framing [4].

In our initial implementations, each layer in the stack had two queues that held message blocks coming in from either above or below. Inter-layer communication was accomplished by moving blocks into these queues and by using the Windows *Event* inter-process communications mechanism to signal their arrival. Performance benchmarks showed that the time taken in queue-manipulation was significant and these were abandoned in favour of using Windows *Messages* that yielded a considerable performance improvement.

In order to preserve modularity, each layer should have minimum coupling with its neighbours. As part of the stack assembly process, a layer is *pushed* onto the top of an (possibly null) existing stack. This process allows a component approach to be taken both to stack development and assembly. If necessary, layers providing different functionality can be added dynamically.

3. REALIZING THE 4TH GENERATION NODE

3.1 General Characteristics of the 4th Generation Node

The wireless node of the future will be flexible, adaptive, intelligent, capable of functioning within standard and proprietary networks, capable of functioning in structured and Ad-hoc environments and suitable for use with multimedia applications. The layered stack architecture is an essential component that enables realization of this 4th Generation Node.

The dynamic layered structure enables a communication stack to be constructed from a suite of protocols. Different routing protocols, transport protocols, media access control (MAC) protocols and Radio interfaces etc. can be inserted and removed as desired. The criteria for choosing the components of the communication stack are varied.

– The communication stack can be created to suit the underlying network or can be adapted to changes in the underlying network. For example, if the medium of physical communication changes from infrared to packet radio the MAC layer module that handles the infrared medium can be removed and replaced it with a MAC layer module that handles the radio packet medium.

– The communication stack can be configured to suit the size and topology of the network. For example, in the case of Ad-hoc networks the most appropriate routing protocol can be changed to suit the number of nodes in the network or the level of mobility of those nodes.

– The communication stack can be configured to suit the underlying communication conditions. For example, a group of nodes in a wireless network could decide to use a more robust but complex channel encoding scheme under noisy communications conditions at the expense of processing speed.

– The communication stack can be configured to deal with non-homogenous networks. For example, to ensure that all nodes can communicate a configuration based on the capabilities on the most primitive node in the network can be selected.

– The communication stack can be configured to suit the application. For example, the chosen transport protocol can be altered for real time communications.

– The communication stack can be configured based on security issues. This can be as simple as choosing a particular kind of encryption for use in the wireless group to as complex as changing the whole communication stack structure if under threat of infiltration.

– The communication stack can be configured based on payment or Quality of Service (QoS). Certain layers of the communication stack may be 'cheaper' than others.

Whatever the chosen configuration, the appropriate layers are placed in each node of the network and assembled at runtime to form a complete protocol stack.

3.2   Wireless Links

The use of the layered architecture makes it easy to experiment with a variety of different wireless links. Conceptually, a radio layer simply accepts packets from the layer above and transmits these packets to every node within range. We have constructed such layers to support infra-red links, short-range UHF radio and also Bluetooth [5] spread spectrum links.

In the Infra-Red case, we envisaged a scenario where a public space could be equipped with a number of infra-red access points to which mobile nodes could 'dock' whenever they needed to connect with other nodes and services on the fixed network or with other mobile nodes that were in a 'docked' state themselves. Transmission is very simple in this scenario as there are typically only two parties involved.

More mobile communications can be achieved using our UHF short-range radios. These make use of an FM radio module on amateur frequencies giving data rates of approximately 64Kbps over distances of 100 metres or so. These radios were chosen to allow full control over what is transmitted and what MAC algorithm is adopted.

More recently, we have additionally begun to use radios based on Bluetooth technology that offer rates of up to 1Mbps over short ranges, but can only do so in the context of a picocell constructed using control dialogues. Packets sent on a Bluetooth link are directed at a destination node, and thus this is one of the parameters that should be added to the 'blackboard' associated with each message block that is transmitted. The layer interface hides many of the Bluetooth idiosyncrasies from higher levels and allows us to construct real Ad-hoc networks consisting of multiple hops over such radio links.

There are multiple possibilities for realising the physical wireless link. To date, in our implementation, options are available for choosing Infra-red (IrDa), Bluetooth, and proprietary UHF radios as the frontend hardware for the system. The generic nature of the stack however means that interfaces to other hardware front-ends are possible.

To extend the size and scope of the network system beyond the number of available physical nodes, a means of simulating an unlimited number of extra nodes has been designed. A *datagram layer*, based on the use of sockets, enables the simulation of physical media such as radio broadcasting and IrDA. This layer offers the same interface to upper levels as a wireless link, but communication actually takes place using IP datagrams. Configured into each node's datagram layer is a list of IP addresses of nodes that are within the simulated radio range. A frame that is sent downwards through such a stack will be transmitted as a datagram to each of the listed nodes just as if it were broadcast on a radio channel.

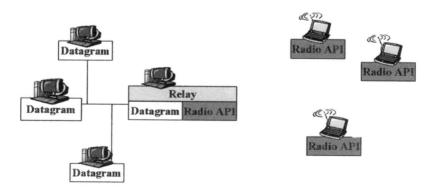

Figure 2. Wireless nodes connected to a fixed access point

We have also constructed a 'relay' layer that can interconnect nodes with real radio interfaces together with a population of nodes that are just connected to the fixed network and making use of the simulated radio layer as illustrated in Figure 2. In this figure wireless nodes can communicate with a fixed network via an access point that is configured with a radio layer, a relay layer and a datagram layer. Note the higher layers of the communication stack on each of the nodes are not shown in Figure 2. The access point allows large populations of nodes to be networked together, some of which are real wireless nodes and some which are not This assists greatly with the development of upper level protocols and also serves as an easy means to give wireless nodes access to resources only available in the fixed network.

The simulation of the radio broadcast environment inherent in our datagram layer does not allow us to simulate mobility adequately. The list of nodes that receive a given transmission correctly is fixed and the reception is either total or non-existent. In order to improve on this, we have begun development of a reality emulator that allows us to experiment with different mobility scenarios.

The reality emulator works by replacing the radio layer in a stack with an emulator layer. The emulator layer in each node connects via a stream socket to a common server. A schematic of this system is shown in Figure 3.

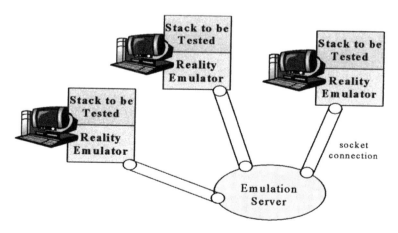

Figure 3. Reality emulator

The transmissions from all nodes are fed via these streams to the server which then emulates the transmission characteristics of the broadcast medium in real-time.

The server can be fed with a map representing a given area and it then places each emulated node at a point on this map. Figure 4 shows 8 mobile nodes positioned against a map of Trinity College showing circles indicating their transmission ranges.

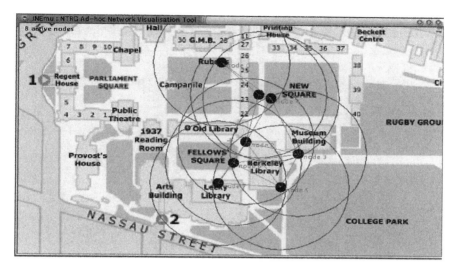

Figure 4. Reality emulator screen dump

Mobility scenarios involving node direction and speed can be input to the system and repeatedly played. Based on the relative positioning of nodes, the server's simple model of the radio channel can cause nodes not to hear each other, for transmissions to collide and for error rates to rise. Since the emulator is configured to emulate our relatively low-speed UHF radios, and the model of the radio channel is quite simplistic, a single server is quite capable of emulating a large number of nodes simultaneously in real-time.

The layers that sit on the reality emulator are unchanged from those version that run on the real radio layers, and the emulator simply makes it much more convenient to exercise them in a wide variety of mobility scenarios.

3.3 Software Radio Layers

Typical radio front-ends perform baseband processing and IF and RF communication functions. In the last five years interest in designing re-programmable radios has increased and software defined radio employing adaptable hardware devices such as Field Programmable Gate Arrays (FPGAs) and Digital Signal Processors (DSPs) are being designed to enable multiple radios to be implemented on a single hardware configuration.

To truly realise a flexible 4th generation wireless node we believe that adopting a general-purpose processor software radio approach gives us more flexibility when it comes to linking services and communication systems. In the long-term, with increasingly powerful CPUs and with the availability of instruction set extensions such as Intel's MMX and SIMD that much of the software radio functions can be carried out using a general-purpose CPU and thus minimise the need for external radio-related hardware. As the Win32 operating environment has a native implementation of threads available on all relevant platforms the multitasking inherent in implementing the software radio functionality can be easily represented in code.

In the software radio case, the radio API layer of the dynamic stack is replaced with a set of software radio layers. This consists of a sequence of 'transformer' layers that will each carry out a single well-defined signal processing operation on the data passing through.

The possibility of being able to specify the baseband processing functions as well as the higher layers of the communication stack means that a truly flexible and adaptable wireless node can exist.

3.4   Ad-hoc Networking Protocol Layers

The adaptable node of the future will often be required to operate in an Ad-hoc environment. Ad-hoc routing protocols seek to address the fundamental problems of route discovery and data trafficking in mobile networks. As Ad-hoc networks are subject to unplanned growth, reduction or fragmentation, any routing algorithms employed must be able to respond to the changing nature of the network's topology. Changes to the network topology manifest themselves in many ways; link cost, link symmetry, route congestion, node mobility, and node density to name a few. These dynamic attributes of an ad-hoc network must be accounted for in the design of any successful routing protocol [6].

Under the stewardship of the IETF.'s Mobile Ad-Hoc Network (MANET) working group, various routing protocols have been proposed and are at different stages of analysis and development [7]. These protocols fall into the two broad categorizations of proactive and reactive protocols. Proactive, table-driven protocols attempt to keep an up-to-date view of the network topology, and in the main are adapted and improved versions of the link-state and distance-vector approaches taken to routing in traditional wired high-bandwidth networks. Reactive protocols such as the Dynamic Source Routing (DSR) [8] protocol will only seek out a specific route when the source requires it. More recent approaches have sought to leverage the optimal aspects of both reactive and proactive protocols. One example is the Zone Routing Protocol (ZRP) [9], a hybrid protocol that incorporates both proactive and reactive elements. In the ZRP, nodes proactively maintain information about neighbouring nodes up to a certain distance way-the zone radius. If routes to destinations outside this zone are sought, then a reactive protocol is employed to dynamically query nodes lying outside the node's zone.

Standard Ad-hoc routing protocols (reactive, proactive and hybrid) have been implemented as layers of the dynamic communication stack and can be tested on real networks [10], can be tested using the simulation facilities and can be tested using the Reality Emulator.

The work in this area has shown that it is clear that the creation of a routing protocol that suits all types of environments and operating conditions is not possible as different protocols excel in different scenarios. To this end adaptive routing protocols are being designed.

Our adaptive approach works on two main levels. Firstly our approach involves creating a system that can facilitate the choice of the best protocol (from a suite of available protocols) to suit the constraints of the scenario. Secondly the adaptive approach also extends to a sub-protocol level, i.e. the protocols themselves could have adaptive elements. To be able to deal with this level of adaptability, the system must has the following characteristics:

1. The system must have the capability to select the most appropriate protocol or change from one routing protocol to another based on stringent decision-making rules. The decision making process must be robust and dynamic.

2. Within any given protocol there are a wide number of controlling parameters. The system must have the ability to optimise individual protocols in a reliable fashion. The optimisation can be local or network wide.

3. The decision-making process must be dynamic. For example when a wide range of information is available to a node, then complex decisions can be taken. When less information is available (for example due to hardware failure, or due to a security alert that makes information from other nodes suspect) decisions must be made using whatever information is available. In the worst- case scenario a base level routing protocol must be used.

4. The system must have a mechanism for communicating the resulting decisions (which may also result in a change of the underlying operating parameters of the communication system) to the rest of the network.

Such an *adaptive Ad-hoc routing system* can effectively deal with the routing issues and problems arising in a hostile and changing environment.

The flexible layered architecture approach is a key component of this system.

3.5   Network Applications

One of the earliest application layers that we developed was designed to allow Internet access from the mobile nodes. Since all of our mobile nodes (both laptop and Palmtop) already run conventional Internet browser software, we were able to connect this to our protocol stack using the concept of a web proxy server. The application layer in the mobile node listens on a local socket for web requests from a suitably configured locally running web browser. The lower layers in the protocol stack are then used to deliver packets from this application layer to a corresponding proxy-server layer running on a node with fixed network access. This then issues the web requests to web servers on the Internet and forwards the results back to the mobile node. This allows us to easily deliver web access to a population of nodes in a wireless Ad-hoc network.

We see point-to-point voice communications as a key application for 4th generation mobile nodes. End-to-end voice transfer coupled with the ability to locate users as they move from node to node on the fixed network, or as their wireless node moves into and out of range of various fixed network access points will be essential if 4th generation systems are to take over the role that mobile phones occupy today.

The basic audio capability is provided in the layered architecture by a general-purpose layer that captures audio and sends it down through the protocol stack. Audio packets coming from below are played to the hardware device. When this layer is coupled with an application layer that implements a signalling protocol such as the Session Initiation Protocol (SIP), we have the basis for a mobile telephony system. The issue of handover is not addressed as yet.

The flexible layered architecture can be also exploited to enhance the performance of multimedia communications. For example, to enable multimedia applications over Ad-hoc networks the issue of error resilient transmission must be addressed. We have designed a new error detection and concealment technique to facilitate error resilient transmission [11]. This technique exploits information from the decoded image data itself and also uses information from the underlying network. The information from the underlying network is accessed by inserting appropriate layers in the communications stack to pass information of interest to the application layer. When applied to MPEG4, the method can localise errors to an even greater extent than with reversible variable length codes (RVLCs) alone.

4. THE 4TH GENERATION SECURITY ARCHITECTURE

Wireless networks, and Ad-hoc networks in particular, derive part of their appeal from the fact that they enable all sorts of communication that was not previously possible. Such openness brings increased security risks. Where a very large number of nodes share the same radio space and are coming into and out of contact with the fixed network, it will become very important to ensure that parties are authenticated before being allowed to communicate with each other and to access shared resources. In cases where the fixed network is accessible, it will be possible to check back to a central server to verify the good standing of a node, but in a disconnected Ad-hoc scenario compromises may have to be made in security to ensure the proper functioning of the Ad-hoc cluster of nodes.

Nodes in our 4th generation system will authenticate themselves on the basis of a multi-faceted identity. At a very basic level, if a node can perform a simple friend-or-foe identification on a neighbour, it may be content to route packets through that node. Authentication of a much higher level may be required before they will enter into an application level dialogue involving sensitive information. Nodes can take on multiple identities simultaneously and these identities may also be hierarchical in nature. When authenticating, a node can progressively reveal more and more of their identity. We represent the different facets of our user's identity as a set of property based digital credentials.

Each node in the system has a credential exchange agent modelled on the work of Winsborough [12], which is invoked whenever an application in this or another node requests access to a *service*. Figure 5 shows a schematic of the credential exchange agent. If the service request is coming from a remote node, the Service Policy will determine which credentials are necessary to allow it to be accessed. For example, access to the service that allows a node to route packets for neighbours in an Ad-hoc network may only require a credential proving that the requester is affiliated with the same employer. A request to be included in a particular multicast audio distribution, on the other hand, may require that user to produce a much more tightly constrained credential proving that the requester is a member of the authorized group of receivers.

In normal human interaction, people do not blindly reveal all of their diverse forms of identification (e.g. drivers licence, passport, employee I.D. card etc) to each person they meet. Often, the person wishing to inspect a credential may need to have some facet of their identity verified before the credential can be handed over. For example, a person may be willing to hand over medical records only to someone who has proved to be a member of the medical profession. Similarly, nodes in our system engage in a progressive credential exchange revealing only those credentials that are required to access the requested services. Credential release policies may specify the conditions that must be satisfied before a sensitive credential may be released.

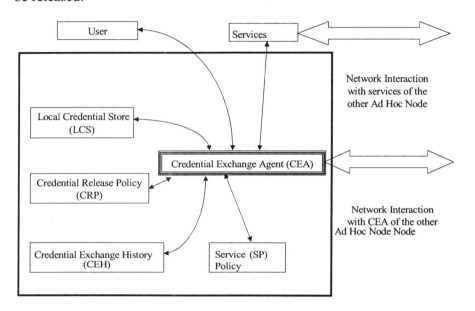

Figure 5. The credential exchange agent

Based on the identities established, multiple nodes will bind together to form groups. A node may be a member of several groups simultaneously. An example to illustrate this may involve multiple emergency services arriving at the scene of an accident. All nodes will join a basic group allowing them to route traffic for the others maximizing connectivity. Paramedic and fire-fighting personnel will each form sub-groups that use this connectivity to have end-to-end dialogues with each other without the possibility of crossover between the dialogues. One can also imagine that other groups could be present in the same space without the majority of nodes even being aware of their existence.

In the Internet multicast community, much work has been done to facilitate the dynamic formation of groups of users that allows users to join and leave these groups while preserving perfect forward and backward secrecy. An example of such system is the Versakey [13] system developed at the ETH in Zurich, Switzerland. These systems tend to rely on the fact that a single node takes on the role of overall controller of the system. In a highly dynamic wireless environment, it will be necessary to form groups in circumstances where this is not possible. It will also be necessary to create new groups on-the-fly and to pass the group leadership role from one node to another.

Our 4th generation security architecture also addresses the issue of resource usage. The co-operative environment of Ad-hoc networking relies on a node's willingness to relay traffic for its neighbours. This activity will deplete a node's battery and also potentially reduce the bandwidth available for its own transmission. One way to address this problem is for nodes to agree only to relay traffic for nodes that are part of an identified group and then to tightly restrict the membership of that group. A more flexible way to address resource usage is to find a means by which nodes can 'pay' each other in real time for any resources consumed.

We have taken the concept of a micropayment, a lightweight cryptographic technique for making repeated payments of very small amounts, and extended it to work with multiple parties [14]. This allows a node to communicate end-to-end using the services of many intermediate nodes. A single stream of payment tokens is injected at the source interspersed with normal data. Conceptually, this represents a single payment that passes through all nodes along the path. Each of the nodes then 'breaks-off' part of the payment commensurate with its contribution to maintaining the end-to-end link. We can use this to compensate mobile nodes for the use of their resources and also to reward organizations that provide wireless access points to the fixed network. This technology essentially enables anyone who has access to the fixed network to in effect become a network operator. The fact that nodes do not have to sign-up in advance with a single organization should allow for increased competition in networks of the future.

Reliable authentication, flexible and secure group formation capabilities and a secure means of payment for resource usage are key elements of the 4th Generation security architecture.

5. SUMMARY

The ultimate goal of our work is to produce a general-purpose mobile node capable of running a large range of network applications that can adapt its mode of operation to the radio environment and prevailing wireless network architecture. To achieve this goal we have created a software and hardware architecture that allows us to experiment with many different facets of wireless networking. We have experimented with multiple wireless links, MAC protocols and Ad-hoc routing protocols. We explored security issues and designed multimedia applications for the system. The resulting node architecture is one that embraces flexibility and adaptability at all levels from the hardware front-end (through software radio) to the application layer and is also an architecture that facilitates multiple levels of security.

REFERENCES

[1]   D. O'Mahony, "Universal Mobile Telecommunications Systems: The Fusion of Fixed and Mobile Networks," *IEEE Internet Computing*, vol. 2, no. 1, January/February 1998, pp. 49-56.

[2]. ISO, *Information Processing Systems - Open Systems Interconnection - Basic Reference Model*, 1984, ISO-7498.

[3]   D.M.   Ritchie, "A Stream Input-Output System," *Bell System Technical Journal*, vol. 63, no. 8, 1984, pp. 1897-1910.

[4]   D. D. Clark and D.L. Tennenhouse, "Architectural Considerations for a New Generation of Protocols," *Proc. of the SIGCOMM Symposium on Communication Architectures and Protocols* (September 1990). *Computer Communication Review*, vol.20, no.4, Sept. 1990.

[5]   *Bluetooth Technical Specification*, Volume 1 Core, Volume 2 Profiles, www.bluetooth.com

[6]   Z.J. Haas and S. Tabrizi, "On Some Challenges and Design Choices in Ad-hoc Communications," *IEEE MILCOM'98*, Bedford, MA, October 18-21, 1998.

[7]   R. Elizabeth and C-K Toh, "A Review of Current Routing Protocols for Ad-hoc Mobile Wireless Networks," *IEEE Personal Communications Magazine*, April 1999, pp. 46-55.

[8]   D. Johnson and D. Maltz, "Dynamic Source Routing in Ad-hoc Wireless Networks," *Mobile Computing*, edited by T. Imielinski and H. Korth, chapter 5, Kluwer, 1996, pp. 153-181.

[9]   Z.J. Haas, "The Routing Algorithm for the Reconfigurable Wireless Networks," *ICUPC'97*, San Diego, CA, October 12-16, 1997.

[10] T. Forde, L.E. Doyle and D. O'Mahony, "An Evaluation System for Wireless Ad-hoc Network Protocols," *Irish Signals and Systems Conference*, June 2000, pp 219-224.

[11] L.E. Doyle, A. Kokaram, and D. O'Mahony, "Error-resilience in Multimedia Applications over Ad-hoc Networks," *Proc. of ICASSP*, Salt Lake City, Utah, May 2001.

[12] W. Winsborough, K. Seamons and V. Jones, "Automated Trust Negotiation," *DARPA Information Survivability Conference and Exposition*, Hilton Head Island, SC, January 2000.

[13] M. Waldvogel, G. Caronni, D. Sun, N. Weiler and B. Plattner, "The VersaKey Framework: Versatile Group Key Management," *IEEE Journal on Selected Areas in Communications*, Special Issue on Middleware, 17(8), August 1999.

[14] M. Peirce and D. O'Mahony, "Flexible Real-Time Payment Methods for Mobile Communications," *IEEE Personal Communications*, vol. 6, no. 6, December 1999, pp. 44-55.

Chapter 5

INTERFERENCE AVOIDANCE USING FREQUENCY LOOKAHEAD AND CHANNEL HISTORY

Supratim Deb *, Manika Kapoor, Abhinanda Sarkar
Coordinated Science Lab, University of Illinois at Urbana-Champaign, USA; IBM India Research Lab, Indian Institute of Technology, New Delhi, India; IBM India Research Lab, Indian Institute of Technology, New Delhi, India

Abstract: Channels in wireless communication systems can only be used during those periods of time when the mobile devices have an interference free connectivity with the central base station. We propose a channel history based scheme in which the base station – a Master for a Master-Slave kind of system – tries to predict whether transmission to a particular user device on a particular frequency will encounter errors due to interference. Our scheme is particularly suited to wireless systems which employ a fixed frequency hopping sequence for transmission and hence cannot use adaptive frequency hopping techniques reported in prior literature. We demonstrate the advantage of the scheme on top of a Master driven frequency hopping system derived from the Bluetooth specification [1]. The scheme prevents transmission on channels seeing bad connectivity and thus insulates the wireless system from interference in the environment. Our simulation results show that using the proposed scheme gives substantial improvement in the throughput and goodput of the wireless system and that it performs better when the wireless environment is more error prone. We have analyzed the scheme using Markov chains. Numerical results from analysis show the performance of the scheme. We show that, with the right tuning of parameters, we can achieve high accuracy in identifying the error-free and error-prone time periods on the wireless channel.

Keywords: Bluetooth, Channel history, Interference, Markov chains, Master-Slave, Two state model

*This work was completed when the first author was working with IBM India Research Lab.

1.   INTRODUCTION

Bluetooth [1,2] and several other indoor wireless networks operating in the Industry Scientific and Medical (ISM) band (2.45GHz $\pm$ 0.05 GHz) use frequency hopping to combat the problem of interference. Common sources of interference could be devices compliant with the Home-RF or the WLAN (IEEE 802.11) standards which also operate in the same band. Also, water molecules resonate at microwave frequencies around 2.4GHz. In the operational environments of super markets, warehouses and factory floors there can be many objects of water content (such as racks of water bottles, fountains etc.) in the vicinity of communicating Bluetooth units. These can absorb much of the ISM-band radiation and reduce the signal strength considerably. Therefore the problem of devising a technique to insulate wireless devices from surrounding interference is a relevant one in context of the pervasive computing paradigm of the future.

Both Frequency Hopping ([3] by Gluck et al., [4] by Bark et al., [5] by Ywh-ren et al.) and Adaptive Frequency Hopping ([6] by Wong et al., [7] by Andersson et al.) provide good immunity against randomly occurring interference signals of short duration. In many applications, however, persistent errors of much larger duration are the likely source of problems. Such errors are caused by electromagnetic emissions from other co-located devices like copy-machines, printers, microwave ovens, baby monitors, etc. These devices stay operational for durations ranging from 10-20 minutes to 2-3 hours and are expected to cause errors due to interference for the entire duration. This interference degrades the throughput of the system considerably even when FHSS (Frequency Hopping Spread Spectrum) techniques are employed. Furthermore, adaptive techniques requiring changing the hopping sequence to avoid bad frequencies are not feasible for systems like Bluetooth which define a fixed hopping sequence for the duration of the connection.

Inoue et al. use channel state information for resource scheduling in [8,9] and [10], where estimation of the channel state has been done according to the successful or unsuccessful receipt of the last slot of a frame. However, the loss of one Medium Access Control (MAC) layer ACK (acknowledgement) does not necessarily mean that persistent interference has set in. In [11], by Fantacci et al., the status of the channel is judged based on the series of acknowledgements from the receiver. However, their policy attempts transmissions at the beginning of every service period, which will be wasteful if the bad period duration is long.

In [12], Bhagwat et al. propose a Link State Monitor (LSM) which assumes perfect channel knowledge and marks/unmarks the relevant Slave queues at the start/end of a burst duration. They then define several Channel State Dependent Packet (CSDP) scheduling policies, all of which rely on the LSM to give correct information about the good/bad state of the link. It is proposed that the physical layer send detailed packet reception status, like CRC-success/failure and power of reception information, to the LSM to help it distinguish between the two states of the channel, but details of the link state monitoring process are not given.

In the same spirit, in [13], Fragouli et al. propose a modified Class Base Queuing model using a Channel State Determiner (CSD), where the CSD estimates the value of a link-goodness parameter. The parameter value corresponds to the maximum number of RTS-CTS (Request To Send, Clear to Send) retries allowed on the corresponding link. If that many tries result in failure, the parameter value is decreased exponentially while if success occurs, its value is increased in inverse proportion to the number of RTS-CTS retries required for the success. Such an iterative technique might take several rounds before it settles down to a good value of the parameter and will require the Master to transmit at least once to the Slave, even if the number of retransmissions allowed is zero. Both these things are wasteful and will possibly lead to poor channel utilization. In [14] by Choi et al., the decision to transmit problem is formulated as a Markov Decision Process and based on this an optimal (randomized) policy for suspending/resuming transmissions is devised which minimizes the cost associated with resource waste due to failed/no transmissions.

It is clear that the performance of these and any other channel state-dependent wireless resource management policy will heavily depend on the accuracy of the scheme used to predict the channel state. We present an easy to implement methodology for taking the decision to suspend/resume transmissions, which can be used in conjunction with CSDP scheduling. We have analyzed it for the case when the error process is Markovian but our simulations show that the methodology works well independent of the model of the underlying error process.

This chapter is based on work [15] by Deb et al. published in IEEE Infocom 2001. Here, as in [15], we extend [16] by Anvekar et al. and propose a method which uses the history of errors seen on the channel by previous packet exchanges, (called the Link State History (LSH)) to predict the state of the wireless channel between the Master and each of its Slaves at every scheduling instant and to decide whether or not to transmit to a particular Slave on a particular frequency. We have modeled the channel as a two state process. We use elementary stochastic processes to analyze the working of the LSH scheme and determine thresholds which estimate the start and end of the bad period. We have carried out detailed simulations to show the correctness of the analytical results and we also provide performance improvement percentages in terms of the throughput and the goodput of the channel.

The rest of this chapter is organized as follows. In the next section we give a brief overview of Bluetooth. In Section 3 we discuss the system architecture, the channel model and spell out the problem statement clearly. In Section 4 we present the LSH scheme and in Section 5 we show analysis of the scheme using Markov chains. Numerical and simulation results, showing the accuracy of this scheme and goodput/throughput improvements, are given in Section 6 and our conclusions are given in Section 7.

2.   OVERVIEW OF THE BLUETOOTH SYSTEM

Given below is a brief description of Bluetooth. More details may be found in published literature, such as reference [2].

The Bluetooth specification defines a low-cost short-range radio link, providing a universal bridge to existing data networks, a peripheral interface, and a mechanism to form small private Ad-hoc groupings of connected devices away from fixed network infrastructures. The technology has a centralized cellular communication model in which each cell (called a piconet) covers an area of approximately 300 square meters and can have a maximum of eight devices. When in operation, one device will act as the Master of the piconet and the other(s) will act as Slave(s) for the duration of the piconet connection.

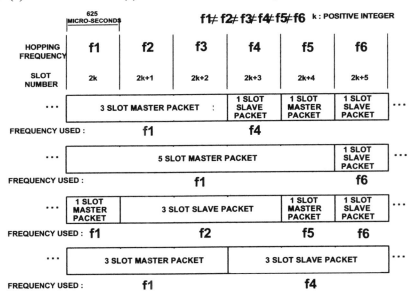

Figure 1.   The Bluetooth channel structure

The Master is required to always start transmission on an even slot and the Slave on an odd slot as indicated in Fig. 1. A Slave is permitted to transmit on a Slave-to-Master slot only if the same Slave was addressed by the Master on the preceding Master-to-Slave slot. It is the Master which determines which Slave is scheduled when and how often. Packets can occupy 1, 3 or 5 time slots. For a multi-slot packet, the hop frequency to be used for the entire packet is the same and is derived from the Bluetooth device clock value for the first slot of the packet. The hop frequency used in the first slot after the multi-slot packet is derived from the current value of the Bluetooth device clock.

Communication takes place over a slotted channel with a nominal slot length of $625\mu s$. The channel is shared among all the devices in one piconet. The channel is represented by a pseudo-random hopping sequence hopping through 79 frequencies in the 2.4 GHz range. Each slot in the channel corresponds to a different hop frequency. The nominal hop rate is 1600 hops/second. The hop frequency corresponding to a particular time slot is determined by a functional unit called the Frequency Selection Unit (FSU) using the Bluetooth device Id of the Master and the value of the Master's clock for the same time slot. The hopping sequence for a piconet cannot be changed as long as the piconet's Master remains the same.

3. SYSTEM ARCHITECTURE, CHANNEL MODEL AND PROBLEM STATEMENT

3.1 System Architecture and Channel Definition

The LSH scheme has been designed for a centralized cellular wireless network architecture. In such a system, several mobile devices (Slaves) maintain point-to-point wireless links with a central base station (Master).

Previous work on CSDP scheduling policies ([12] by Bhagwat et al., [13] by Fragouli et al.) model each Master-Slave link as separate two state Markov processes, but they do not take into account the fact that most current indoor cellular wireless systems use frequency hopping for better performance in congested radio conditions. Experimental work by Zander et al. in [17] shows that the bit error rates seen by different frequencies in the hopping spectrum are quite different from each other. We therefore argue that a logical division of the channel into Master-Slave links is not enough. In order to completely capture the picture of interference present in the environment, each Slave-frequency pair should be considered separately. For this reason, we define one **channel** to be one Master-Slave-frequency link.

3.2 Channel Model

The modeling of indoor wireless channel has been studied extensively in the literature. Reference [12] by Bhagwat et al. makes the observation that frequency hopping WLANs may hop to a frequency which is more susceptible to interference. The indoor channel has been modeled as a two state burst-noise model using Markov chains in [18] by Roberts et al. A two state continuous time model has been proposed by Bucher et al. in [19].

As is done in previous literature, including [13] by Fragouli et al., we model the channel as a two state stochastic process. The channel stays in the good state for an amount of time X_g and goes to the bad state when interference sets in and stays bad for an amount of time X_b. We denote the good and the bad states by 0 and 1 respectively. Unlike [13] by Fragouli et al., however, we do not make the assumption that the good state is always error free, although the bad state is always destructive. Instead, we analyze the LSH scheme using p_g and p_b as the probabilities of receiving a packet correctly in the good and the bad state, where p_g is high (around 0.90 to 0.95) and p_b is low (around 0.10 to 0.05). Let $X(t)$ denote the state of the channel at time t. If X_g and X_b are exponentially distributed with mean λ^{-1} and μ^{-1} respectively, then the process $X(t)$ is a Continuous Time Markov Chain, for which the following can be shown from renewal theory ([20]).

$$
\begin{aligned}
p(t) \;&\overset{\text{def}}{=}\; P(X(t+\tau) = 0 \mid X(\tau) = 0) \qquad\qquad (5.1)\\
&= \; \frac{\mu}{\mu+\lambda} + \frac{\lambda}{\mu+\lambda}\exp(-(\mu+\lambda)t)
\end{aligned}
$$

3.3   Problem Statement

We address the problem of time varying connectivity in wireless communication systems with centralized architecture. At any scheduling instant, the Master should communicate with only that subset of its Slaves which are in the good state. Ideally, the Master should stop transmission on a channel as soon as its state goes bad and resume transmission as soon as it comes back to the good state.

A problem arises because the connectivity of the devices changes unpredictably with time and both the 'bad' and 'good' periods last for intervals of random duration. The task is to devise a methodology to identify the switch from the good to the bad state as early as possible, and also to deduce that the channel has become good at the end of the duration of the bad period. Such a scheme can then be used by the Master to pick the subset of good connectivity devices at each scheduling instant.

If such a scheme recognizes the bad period later than the actual time it starts, then channel utilization will fall, and if it recognizes the start of a good period too late, then the corresponding Slave device will get starved of data even during the good period. So the accuracy of the scheme is crucial in maintaining a good level of channel utilization.

4. THE LSH SCHEME

The LSH scheme is used by the Master to identify devices having a high probability of seeing bad connectivity and avoid communication with them while they are in that state. With the right tuning of thresholds, the scheme achieves good accuracy in recognizing both the beginning and the duration of the bad period. In this chapter, the scheme has been discussed for a Master driven frequency hopping system derived from the Bluetooth specification [1], but its basic approach can be applied to any other wireless communication system having a centralized architecture. The different aspects of the scheme are as follows.

- **Maintenance of a Link-State-History table with the Master:**
  The Link State History (LSH) table, in Fig. 2, has one counter corresponding to each channel (Slave-frequency pair), on which the Master can potentially transmit. The counters record recent history of errors experienced on a particular channel. Two thresholds $T_{TRANSMIT}$ and T_{RESET} are defined for the counter values. The physical interpretation of these thresholds and how they are obtained is explained in the following sections.

	FREQ 1	FREQ 2	FREQ 3		FREQ 78	FREQ 79
SLAVE 1	C(1,1)	C(1,2)	C(1,3)	. . .	C(1,78)	C(1,79)
SLAVE 2	C(2,1)	C(2,2)	C(2,3)	. . .	C(2,78)	C(2,79)
SLAVE 3	C(3,1)	C(3,2)	C(3,3)	. . .	C(3,78)	C(3,79)
SLAVE 4	C(4,1)	C(4,2)	C(4,3)	. . .	C(4,78)	C(4,79)
SLAVE 5	C(5,1)	C(5,2)	C(5,3)	. . .	C(5,78)	C(5,79)
SLAVE 6	C(6,1)	C(6,2)	C(6,3)	. . .	C(6,78)	C(6,79)
SLAVE 7	C(7,1)	C(7,2)	C(7,3)	. . .	C(7,78)	C(7,79)

C(i, j) = COUNTER FOR LINK STATE OF SLAVE_i FOR FREQ_j

Figure 2. The LSH table

- **Maintenance of a Frequency Look-ahead Unit (FLU) with the Master:**
  The proposed algorithm requires a Bluetooth Master unit to be equipped with an additional FSU to find frequencies corresponding to future time slots, by providing as input to this FSU, the device address bits of the Master unit and the appropriate system-clock bits of the Master unit as

shown in Fig. 3. Although look-ahead for sequential time slots is implied in this flowchart, the Master can, in general, look-ahead for future time slots in any arbitrary order by just providing the clock bits corresponding to the required slot.

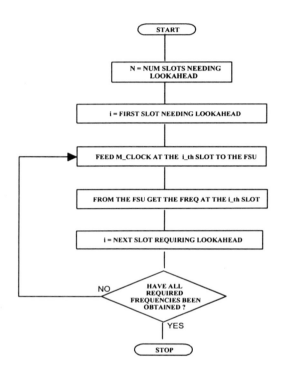

Figure 3. Flowchart for sequential lookahead of N frequencies

- **LSH update based on packets received by the Master:**
  The LSH counters record the most recently seen number of consecutive NAKs (negative-acknowledgements) received by the Master on the corresponding channel. Initially, all the counters are reset to zero. Depending upon whether the current value of the counter is below or above $T_{TRANSMIT}$, there are two ways by which the counter value can be incremented or reset. Consider the case when the Master is communicating with Slave-'i'. Let f_j be the frequency of transmission by the Master and f_k be the frequency on which the Slave replies.

  If the counter $C(i,j)$ is below $T_{TRANSMIT}$ the Master allows transmission/reception to take place to/from Slave-'i' on frequency f_j and changes counter value depending on the success or failure of the trans-

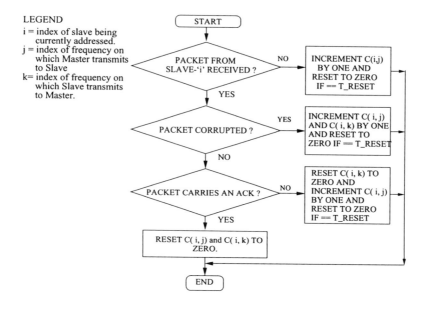

LEGEND
i = index of slave being
   currently addressed.
j = index of frequency on
   which Master transmits
   to Slave
k= index of frequency on
   which Slave transmits
   to Master.

START

PACKET FROM SLAVE-'i' RECEIVED ? — NO → INCREMENT C(i,j) BY ONE AND RESET TO ZERO IF == T_RESET

YES

PACKET CORRUPTED ? — YES → INCREMENT C(i, j) AND C(i, k) BY ONE AND RESET TO ZERO IF == T_RESET

NO

PACKET CARRIES AN ACK ? — NO → RESET C(i, k) TO ZERO AND INCREMENT C(i, j) BY ONE AND RESET TO ZERO IF == T_RESET

YES

RESET C(i, j) and C(i, k) TO ZERO.

END

Figure 4. Updating the LSH table

mission. If the transmission was not successful, as indicated by the receipt of a NAK or a garbled packet, the corresponding counter is incremented by 1. If the transmission was indeed successful, as indicated by an ACK, the counter is reset to zero. In Fig. 4 when the Master receives no packet from the Slave it assumes that the Slave did not reply because it never received the transmission due to channel *i-j* being bad, so it increments C(i,j) by 1. If the Master receives a garbled packet from the Slave, it indicates that channel *i-k* was bad. Then, the Master assumes that the packet was carrying a NAK and, therefore, increments both C(i,j) and C(i,k). The receipt of a NAK from a Slave indicates that the Master's transmission on channel *i-j* was not received correctly, but the channel *i-k* on which the Slave sent the NAK is good. Therefore C(i,j) is incremented, and C(i,k) is reset. The receipt of an ACK from the Slave sent on frequency f_k reports successful receipt of the data sent by the Master on frequency f_j, and indicates that both the channels, *i-j* and *i-k* are good. Therefore both C(i,j) and C(i,k) are reset to zero. If the counter value is equal to or above $T_{TRANSMIT}$ then the Master does not allow transmission/reception to/from Slave-'*i*' on frequency f_j/f_k. In this case the Master simply increments C(i,j) by 1. Each time the Master increments the counter, it then checks if the counter value now equals

T_{RESET} , and if it does, the Master resets the counter to zero. In other words the increment operation is modulo(T_{RESET}).

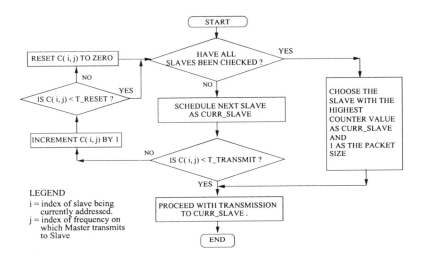

Figure 5.   Slave selection using the LSH

- **LSH table usage for selecting a good connectivity Slave:**
  Again, consider the case when the Master is communicating with Slave-'i' on frequency f_j. The LSH-scheme requires that before transmission (see Fig. 5), the Master consult the LSH table to determine whether the channel (i, j) is in good state or not. If the LSH counter $C(i,j)$ is below a threshold, $T_{TRANSMIT}$, then the Master can go ahead with the transmission. If, on the other hand, the counter value is greater than or equal to $T_{TRANSMIT}$, then the Master is required to not transmit to that Slave and find a different Slave, to which it can transmit at the same scheduling instant (and hence the same frequency), so as not to waste the current slot on the channel. To do so, the Master checks the LSH table for the next Slave, and so on, until it finds a Slave, for which the corresponding counter for f_j is below $T_{TRANSMIT}$. In the event that it can find no Slave with a counter below $T_{TRANSMIT}$, it chooses the Slave whose LSH counter has the highest value. The loss of service by the intended Slave and gain of service by another Slave can be recorded to address the issue of service fairness as per any suitable policy. For wireless systems in which communication happens in slot pairs, transmission is considered unsuccessful if even one of the packets sent on either of the 'to' or 'from' slots get corrupted. For such systems, the Master is required

to use the FLU to obtain the frequency, say f_k, on which the Slave will send its reply, and check both the channels, $i\text{-}j$ and $i\text{-}k$, at each scheduling instant, and carry on with the transmission only if both the counters, C(i,j) and C(i,k), are below $T_{TRANSMIT}$. Note, that using this process, the Master does not ever change the hopping sequence of the wireless system, and hence this scheme can be used for systems with a fixed hopping sequence.

4.1 Physical Interpretation of Counter Thresholds

Before transmission to Slave-'i' on frequency f_j , counter C(i,j) is checked. If C(i,j) $\leq T_{TRANSMIT}$ then this means the channel has recently encountered counter-value-many consecutive errors. As long as C(i,j) $< T_{TRANSMIT}$, the Master allows further transmissions. When C(i,j) equals $T_{TRANSMIT}$, the Master interprets this to mean that the errors are due to persistent interference and backs-off, ceasing transmission on that channel for as long as the bad period persists. After this, every time the scheduling policy determines that the next transmission will be to Slave-'i' on frequency f_j, the Master does not transmit to Slave-'i', instead it simply increments the counter C(i,j) by 1 modulo T_{RESET} to mark the fact that one turn for Slave-'i' on f_j has been missed. Thus C(i,j) between $T_{TRANSMIT}$ and T_{RESET}, means that $C(i,j) - T_{TRANSMIT}$ many turns of Slave-'i' on f_j have been missed as back-off and provides a measure of the amount of back-off time. As long as C(i,j) is below T_{RESET}, it gets incremented at each scheduling instant. When C(i,j) equals T_{RESET} the counter is immediately reset to zero. The next time Slave-'i' is scheduled on frequency f_j, communication can go ahead as the counter was reset to zero the previous time.

The two thresholds $T_{TRANSMIT}$ and T_{RESET} provide an easy way of distinguishing between transient and persistent errors on a link. If a Master-Slave link on a certain frequency is affected with short-term interference, then before $T_{TRANSMIT}$ is exceeded the Master gets an acknowledgment for one of its packets sent to the Slave, and the corresponding counter gets reset. If the interference is persistent, however, then transmissions fail repeatedly. The receiving Slave unit sends NAKs each time this happens. As a result, $T_{TRANSMIT}$ is soon reached and after that $(T_{RESET} - T_{TRANSMIT})$ many transmission attempts are skipped as back-off. The expectation is that the channel will be interference free when it comes out of the back-off period. During this back-off period the Master continues to transmit to other Slaves and also to this Slave on other good frequencies.

5.   PERFORMANCE ANALYSIS OF THE LINK STATE HISTORY BASED SCHEME

We perform a Markov chain analysis to have an insight on the performance of the scheme for different values of the threshold parameters. Let N denote $T_{TRANSMIT}$ and let R denote $(T_{RESET} - 1)$. Consider the counter evolution process at the packet transmission attempt epochs. Let t_k denote the time of kth opportunity (it may be a decision to transmit or to not transmit depending on the counter value, according to the scheme proposed earlier) to transmit at a particular frequency for a particular Slave. Let the counter value be $C(t_k)$ just before the counter update takes place The channel state is denoted by $X(t_k)$ at those epochs. Note that even under the exponential distribution assumption of the good and the bad states, the joint process $(C(t_k), X(t_k))$ is not a discrete time Markov chain (DTMC). However, under the additional assumption that the value of $(t_k - t_{k-1})$ is deterministic, the process $(C(t_k), X(t_k))$ can be modeled as a DTMC. Since the purpose of this section is to have an insight on the performance of the scheme, we assume that $t_k - t_{k-1} = T$. This is exact for systems where an attempt is made to schedule every channel (a particular Slave-frequency pair) at equally spaced intervals. Let α_T and β_T denote the following quantities.

$$\alpha_T = \frac{\mu}{\mu+\lambda} + \frac{\lambda}{\mu+\lambda}\exp(-(\mu+\lambda)T) \tag{5.2}$$

$$\beta_T = \frac{\lambda}{\mu+\lambda} + \frac{\mu}{\mu+\lambda}\exp(-(\mu+\lambda)T) \tag{5.3}$$

From (5.1), it is clear that α_T denotes the probability that the channel is in good state at time t_k, conditioned upon the fact that the channel was in good state at t_{k-1}. β_T denotes the corresponding probability for the bad state. We now describe the transition probabilities of the DTMC.

Let $p_{(i,j)\to(i',j')} = Prob(A_k|A_{k-1})$.
where A_k is the event $\{(C(t_k), X(t_k)) = (i',j')\}$.
and A_{k-1} is the event $\{(C(t_{k-1}), X(t_{k-1})) = (i,j)\}$.

These transition probabilities are straightforward. Consider the transition from $(i,0)$ to $(0,0)$ state for $i < N$. This happens if the attempted transmission is successful and the channel does not change state. This happens with a probability $p_g\alpha_T$. The probabilities for the other possible transitions can be observed in a similar manner. Note that there is no attempt of transmissions after $C(t_k)$ reaches N and hence it can only increase by 1, until $C(t_k)$ reaches R. The counter is reset to 0 on reaching R. More formally,

$P_{(i,j) \to (i',j')} =$

$$
\begin{cases}
\text{for } i < N, \, j = 0 \,; \\
p_g \alpha_T, \quad \text{if } i' = 0, \, j' = 0 \\
(1 - p_g) \alpha_T, \quad \text{if } i' = i+1, \, j' = 0 \\
p_g (1 - \alpha_T), \quad \text{if } i' = 0, \, j' = 1 \\
(1 - p_g)(1 - \alpha_T), \quad \text{if } i' = i+1, \, j' = 1 \\[2mm]
\text{for } N \leq i \leq R, \, j = 0 \,; \\
\alpha_T, \quad \text{if } i' = (i+1) \bmod (R+1), \, j' = 0 \\
(1 - \alpha_T), \quad \text{if } i' = (i+1) \bmod (R+1), \, j' = 1 \\[2mm]
\text{for } i < N, \, j = 1 \,; \\
p_b \beta_T, \quad \text{if } i' = 0, \, j' = 1 \\
(1 - p_b) \beta_T, \quad \text{if } i' = i+1, \, j' = 1 \\
p_b (1 - \beta_T), \quad \text{if } i' = 0, \, j' = 0 \\
(1 - p_b)(1 - \beta_T), \quad \text{if } i' = i+1, \, j' = 0 \\[2mm]
\text{for } N \leq i \leq R, \, j = 1 \,; \\
\beta_T, \quad \text{if } i' = (i+1) \bmod (R+1), \, j' = 1 \\
(1 - \beta_T), \quad \text{if } i' = (i+1) \bmod (R+1), \, j' = 0
\end{cases}
$$

Since the DTMC is finite irreducible, the steady state probability distribution can be obtained by solving for the left eigenvector of the transition probability matrix. Let $\pi_{(i,0)}$ denote the steady state probability of being in state $(i,0)$ and $\pi_{(i,1)}$ denote the steady state probability of being in state $(i,1)$. The balance equations can be written as follows.

$$
\begin{aligned}
\pi_{(0,0)} &= p_g \alpha_T \left(\pi_{(0,0)} + \pi_{(1,0)} + \cdots + \pi_{(N-1,0)} \right) + \\
&\quad p_b (1 - \beta_T) \left(\pi_{(0,1)} + \pi_{(1,1)} + \cdots + \pi_{(N-1,1)} \right) + \\
&\quad \pi_{(R,0)} \alpha_T + \pi_{(R,1)} (1 - \beta_T) \\
\pi_{(i,0)} &= (1 - p_g) \alpha_T \pi_{(i-1,0)} + \\
&\quad (1 - p_b)(1 - \beta_T) \pi_{(i-1,1)} \,, \text{ for } i \leq N \\
\pi_{(i,1)} &= (1 - p_b) \beta_T \pi_{(i-1,1)} + \\
&\quad (1 - p_g)(1 - \alpha_T) \pi_{(i-1,0)} \,, \text{ for } i \leq N \\
\pi_{(i,0)} &= \alpha_T \pi_{(i-1,0)} + \\
&\quad (1 - \beta_T) \pi_{(i-1,1)} \,, \text{ for } N < i \leq R \\
\pi_{(i,1)} &= \beta_T \pi_{(i-1,1)} + \\
&\quad (1 - \alpha_T) \pi_{(i-1,0)} \,, \text{ for } N < i \leq R
\end{aligned}
$$

These $2R+1$ equations along with the additional constraint that the $\pi_{(i,j)}$s should add to 1, can be solved numerically to obtain the different performance measures.

5.1   Performance Measures

We are interested in obtaining different measures indicating how well the scheme works. Let $P(B \mid No\ Tx)$ be the steady state probability that the channel is in bad state conditioned on there being no attempt of transmission. Correspondingly, let $P(G \mid Tx)$ be the steady state probability that the channel is in good state conditioned on there being an attempt at transmission. These can be defined in terms of $\pi_{(i,j)}$s as follows.

$$P(B \mid No\ Tx) \;=\; \frac{\displaystyle\sum_{i=N}^{R} \pi_{(i,1)}}{\displaystyle\sum_{i=N}^{R} \pi_{(i,0)} + \sum_{i=N}^{R} \pi_{(i,1)}} \tag{5.4}$$

$$P(G \mid Tx) \;=\; \frac{\displaystyle\sum_{i=0}^{N-1} \pi_{(i,0)}}{\displaystyle\sum_{i=0}^{N-1} \pi_{(i,0)} + \sum_{i=0}^{N-1} \pi_{(i,1)}} \tag{5.5}$$

Note that a high value of the first measure indicates how efficient the scheme is whenever there is a decision to not transmit, whereas a high value of the latter measure indicates the efficiency of the scheme whenever there is a decision to transmit. Ideally we would like to maximize both. But we later argue and show in our results that one of these increases with R, but the other decreases.

5.2   Choice of N

The critical parameter in this scheme is R, as we show in our results. Any value of N which is not large should be good enough since the value of N should be such that, once interference sets in backing off is done at the earliest. But since there can be wireless channel errors even during the good state of the channel, it has to be ensured that backing off does not take place when the channel is in the good state. Consider for a moment, the channel to be in good state for a sufficiently long time. Denote by ET_i the expected discrete time units (of the DTMC shown before) for the counter to reach a value of i (for $i \leq N$) starting from counter value 0. We have the following recursion for ET_i in this case.

$$ET_i \;=\; (ET_{i-1}+1)(1-p_g) +$$

$$(2ET_{i-1}+2)(1-p_g)p_g +$$
$$(3ET_{i-1}+3)(1-p_g)p_g^2 \ldots.$$
$$\Rightarrow ET_i = \frac{ET_{i-1}+1}{1-p_g} \; ; ET_0 = 0$$
$$\Rightarrow ET_i = \frac{1-(1-p_g)^i}{p_g(1-p_g)^i} \qquad (5.6)$$

We choose N such that $(ET_N \times T)$ is sufficiently larger than the expected good state duration. If X_g, the duration of the good period is exponentially distributed with mean λ^{-1}, then we choose N, such that the following holds.

$$Prob(X_g < (ET_N \times T)) \; > \; 1-\varepsilon$$
$$\Rightarrow \exp(-(ET_N \times T)\lambda) \; < \; \varepsilon \qquad (5.7)$$

Here ε is a small positive number and can be taken as, say, 0.01. Plugging in the expression for ET_N in the above and solving for N, we get the following.

$$N \; > \; -\frac{\ln(1-\frac{(\ln \varepsilon)p_g}{T\lambda})}{\ln(1-p_g)} \qquad (5.8)$$

N can actually be chosen as the ceiling of the quantity on the right hand side of the above inequality.

6. NUMERICAL RESULTS FROM ANALYSIS, SIMULATION SETUP AND SIMULATION RESULTS

We now show results indicating the performance of the scheme and overall improvement in terms of throughput and goodput. We first show numerical results obtained from analysis, indicating the performance of any particular counter (recall that the decision to transmit or to not transmit depends on the value of the counter). Simulation results are obtained to show that this scheme indeed results in improvement in throughput and goodput.

6.1 Numerical Results from Analysis

Consider the variation of $P(B|No\ Tx)$ and $P(G|Tx)$, given by (5.4) and (5.5), for different values of R. As R is increased for fixed values of N, the back-off duration given by $(R-N+1) \times T$, is increasingly likely to overlap with the next good state duration. If the value of R is too small, then the counter gets reset to 0 too often, and hence although a switch from bad to good state of the channel is detected very fast, a lot of transmissions are wasted as the counter goes from 0 to N while the channel is still bad, more often than it would have for a larger R. Hence there would be a decrease in the value of $P(B|No\ Tx)$ with increasing R and also an increase in the value of $P(G|Tx)$ with increasing R. Since, ideally we would like to have a high value of both, we introduce a quantity called *Utility* as defined below.

$$Utility\ =\ P(B|No\ Tx)\ \times P(G|Tx) \tag{5.9}$$

We plot *Utility* along with $P(B|No\ Tx)$ and $P(G|Tx)$, to study how these vary for different values of R.

Although the analysis is quite general, the value of T in (5.2) and (5.3) has to be chosen in an appropriate manner. We consider the scenario of a Bluetooth piconet which has one Master and at most seven slaves. Hence, we take T to be $7 \times 79 \times 625\mu s = 0.35s$, where there are 79 different frequencies in the hopping pattern and $625\mu s$ is the duration of each slot.

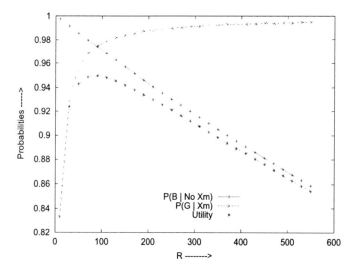

Figure 6.   Curves for $\lambda^{-1} = 1200s$, $\mu^{-1} = 600s$, $p_g = 0.95$, $p_b = 0.05$. The value of N as derived from (5.8) is 4

We show the plots for two sets of p_g and p_b. In one (Fig. 6 and Fig. 7) we take a very small value of p_b as 0.05, for the scenario when the channel is badly affected due to interference. In the other set, (Fig. 8 and Fig. 9) we take p_b to be 0.20. The value of p_g is kept same at 0.95 in both the sets. We have shown the results for two cases, one when the mean bad state duration, (μ^{-1}), and the mean good state duration, (λ^{-1}), are equal and the other when the mean good state duration is larger than the mean bad state duration. Observe that, in all the plots $P(B|No\ Tx)$ decreases with increasing R, whereas $P(G|Tx)$ increases

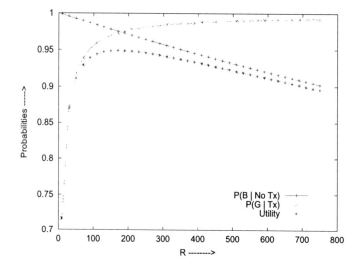

Figure 7. Curves for $\lambda^{-1} = 1200s$, $\mu^{-1} = 1200s$, $p_g = 0.95$, $p_b = 0.05$. The value of N as derived from (5.8) is 4

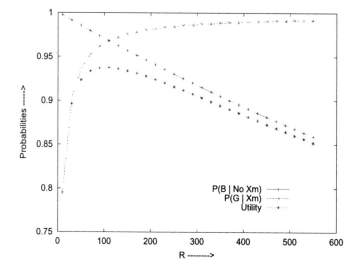

Figure 8. Curves for $\lambda^{-1} = 1200s$, $\mu^{-1} = 600s$, $p_g = 0.95$, $p_b = 0.2$. The value of N as derived from (5.8) is 4

with increasing R as expected and explained earlier in the section. The value of the *Utility* variable as defined in (5.9) initially increases with R and then

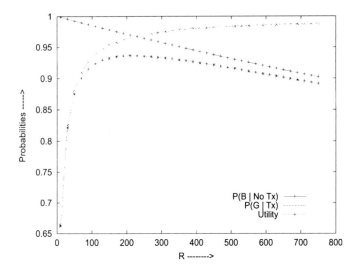

Figure 9. Curves for $\lambda^{-1} = 1200s$, $\mu^{-1} = 1200s$, $p_g = 0.95$, $p_b = 0.2$. The value of N as derived from (5.8) is 4

comes down. Thus, a good design choice of R can be, the value at which *Utility* attains its peak value. For example, when $\lambda^{-1} = 1200s$, $\mu^{-1} = 600s$, $p_g = 0.95$, $p_b = 0.05$, a good choice of R can be 85. In this case, the decisions to transmit and to not transmit, are both correct (in the sense that a decision to transmit actually implies that the channel is good, and a decision to not transmit actually implies that the channel is bad) with a probability of more than 0.95. So, a judicious choice of R, can indeed make the probability of correct decision close to 1. This eventually results in an improvement in goodput, as we show later in our simulation results.

We now make a few more interesting observations. A good choice of R in Fig. 7 is around 175. This is more than the case in which the bad state duration is shorter (Fig. 6). This is intuitively very clear as the backing off duration should be larger when the bad state mean duration is larger. Also, by comparing the curves in Fig. 7 and Fig. 9, we observe that R has a larger value when p_b is larger with everything else remaining same. This can be explained as follows. For a particular value of R, every time the counter comes out of backoff and gets reset while the channel is still in the bad state, note that it takes more time (more transmission attempts) for the counter to reach N again, for the case where the value of p_b is larger. As a result $P(G|Tx)$ is smaller when p_b is larger and the other parameters are same, but $P(B|No\ Tx)$ remains more or less the same. This indicates that the good choice of R should be larger for the larger p_b case, as the analytical results suggest.

We show simulation results with the same set of parameters later in this section.

6.2   Simulation Setup

We have simulated the implementation of the LSH scheme for the Bluetooth communication model. The Master communicates with 7 active Slaves on a slotted channel hopping over 79 frequencies and therefore maintains a table of 7×79 counters, one for each Slave-frequency pair. The pseudo-random hopping sequence is generated using the FSU kernel as defined in the Bluetooth specification [1].

The simulations have been carried out for two sets of (p_g, p_b), these being (0.95, 0.05) and (0.95,0.20). Every Slave-frequency pair is considered separately as a channel. A channel is considered as being either 'clean' or 'interference-affected'. A 'clean' channel does not ever suffer from interference. An 'interference-affected' channel experiences good and bad periods of random duration. Results are presented for the distribution of the good and bad periods being exponential, heavy-tailed, and constant.

Performance evaluation is done using the parameters of throughput and goodput. We define throughput to be the ratio of the number of error-free transmissions between the Master and all its Slaves in a given time interval to the maximum number of such transmissions that the channel can actually accommodate in the same time interval. Throughput does not take retransmissions into account. Goodput is defined here as the ratio of the number of error free message-response pair packets exchanged between the Master and all its Slaves in a given time interval to the maximum number of message-response pair packets that can possibly be exchanged on the channel in the same time interval.

In wireless systems, such as Bluetooth, where communication happens in Master-Slave slot pairs, transmission is considered unsuccessful when even one of the two packets in the pair gets corrupted. In such a case, retransmission of both the packets is required and hence the number of slots wasted are actually two and not one. Goodput takes into account only the amount of meaningful information exchange between the Master and all its Slaves.

We first assess the performance of the scheme using a fixed value of N and R in Fig. 10 and Fig. 11 to show that our scheme provides significant improvement in goodput even if N and R are not chosen in a judicious manner. Fig. 12, Fig. 13, Fig. 14 and Fig. 15 show the variation in the performance of the scheme over a range of values of R.

6.3   Simulation Results

Fig. 10 shows that the improvement in goodput increases as the number of 'interference-affected' channels increases. The simulations have been carried out for a fixed value of N and R, these being 3 and 100 respectively. The LSH scheme prevents transmissions on channels suffering from interference and thus insulates the system from interference in the environment. Our results show that the goodput of the system using the LSH scheme stays at acceptable levels of 60%-90% (90% when 10% channels are affected, and 60% when 100% channels are affected). However, goodput of the base system (i.e. the system which does not use the LSH scheme for error avoidance) falls from 83% to 24%, as the number of 'interference-affected' channels rises from 10% to 100%. So we observe that the LSH scheme can offer improvements as high as 150% and allows the system to remain usable even when all the channels are 'interference-affected'. This can also be explained qualitatively by saying that since the LSH scheme tries to shield the system from bad channels, the more the number of bad channels present, the more will be the improvement seen by the system. Fig. 11 shows the results for $p_b = 0.20$. This shows that the LSH scheme gives 5% to 85% performance improvements when the number of 'interference-affected' channels rises from 10% to 100%. The performance gains in this set of results is lower because p_b, the probability of correct transmission in the bad state is higher at 0.20. Comparison of Fig. 10 and Fig. 11 again goes to illustrate the fact that the LSH scheme gives better performance when the channels are more error prone.

Fig. 12 and Fig. 13 show the variation in the performance of the LSH scheme versus the Reset-Threshold (R) for different interference patterns. For these results we have taken 40% of the channels to be 'interference-affected', p_g to be 0.95 and p_b to be 0.05. The value of N, as derived from (5.8) is taken to be 4. We study three different kinds of interference patterns. The first in which the bad period is constant and the good period has exponential distribution, the second in which both the good and bad periods have exponential distribution, and the third when the good period has exponential distribution and the bad period has a heavy-tailed Pareto distribution (The pdf of this distribution is $f(x) = \frac{\alpha \beta^\alpha}{x^{\alpha+1}}$ and the expectation is $E = \beta \frac{\alpha}{\alpha-1}$ where $\beta \le x < \infty$. We have taken $\alpha = 1.5$ to ensure finite expectation and infinite variance). For each of these patterns, we consider two subcases, one in which the mean duration of the good and bad periods is the same and the other in which the mean duration of the good period is more than the mean duration of the bad period.

The scheme shows a more than 30% goodput improvement and a 15% throughput improvement in all the three cases. Note that, the *improvement* in goodput is always more than the *improvement* in throughput. This is because the LSH-based scheme always checks the health of the next slot as well as the

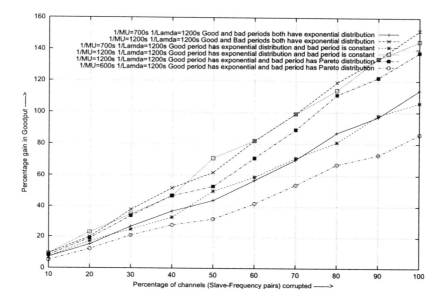

Figure 10. Curves showing percentage gain in goodput vs number of channels corrupted for various distributions of the good and bad period. p_b=0.95, p_g=0.05, N=3, R=100

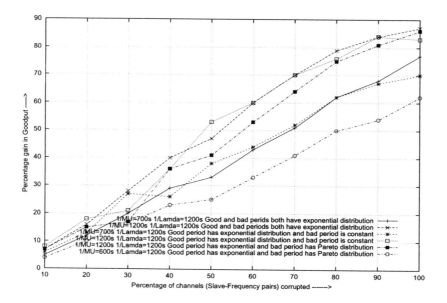

Figure 11. Curves showing percentage gain in goodput vs number of channels corrupted for various distributions of the good and bad period. p_b=0.95, p_g=0.20, N=4

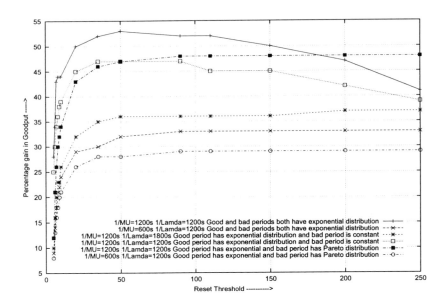

Figure 12. Curves showing percentage gain in goodput vs different values of the Reset Threshold. p_g=0.95, p_b=0.05, N=4

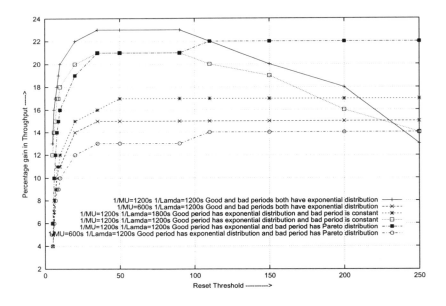

Figure 13. Curves showing percentage gain in throughput vs different values of the Reset Threshold. p_g=0.95, p_b=0.05, N=4

present slot before taking a transmission decision. Hence, the throughput and the goodput are not very different when an LSH based scheme is used and this is corroborated by our simulation results. This is not true when an LSH-based scheme is not there, in which case throughput is interpreted as the long run probability of a slot being utilized and goodput is the long run probability of two consecutive slots being utilized. Degradation in terms of goodput is hence more in the absence of an LSH based scheme. The goodput rises very rapidly for small values of R, plateaus for values of R between 100 and 200, and then starts falling with larger values of R. (Goodput falls for larger values of R in every test-case though this fall is not visible in the range of R we have chosen for the result graphs.) The improvement in goodput is the most (50%) when the ratio of the expected bad period duration to the expected good period duration is the most (1:1) and lesser when the ratio is smaller (1:2). This clearly indicates that the LSH scheme provides very good protection to wireless systems suffering from heavy interference.

There are values of R for which the goodput improvement is the most. Good values of R depend on the mean bad period duration and not on whether the bad period is constant, or has exponential or heavy-tailed distribution, as indicated by Fig. 12 and Fig. 14. This seems to indicate that a good choice of R depends more on the mean bad period duration and less on its actual distribution.

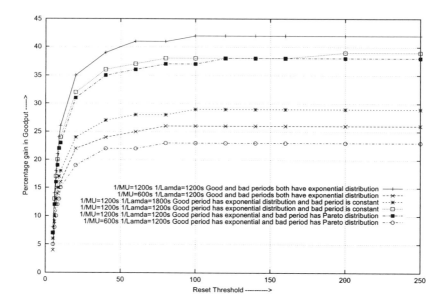

Figure 14. Curves showing percentage gain in goodput vs different values of the Reset Threshold. p_g=0.95, p_b=0.20, N=4

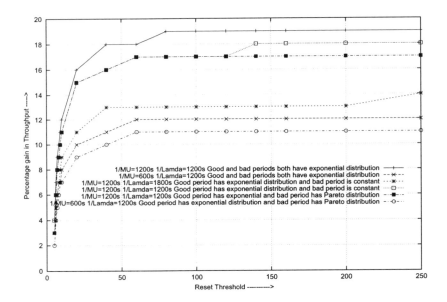

Figure 15. Curves showing percentage gain in throughput vs different values of the Reset Threshold. p_g=0.95, p_b=0.20, N=4

Fig. 14 and Fig. 15 show that the LSH scheme gives best results at a higher value of R when p_b is higher at 0.20, as was seen in the results from the numerical analysis. This indicates that the choice of a correct R is more critical when the channel is less error prone.

7.   CONCLUSIONS

In centralized architecture wireless network systems, the connectivity between the Slaves and the base station varies stochastically between good and bad with time. We have proposed a link state history based scheme which uses the past history of errors encountered on the channel to identify the devices currently seeing bad connectivity and avoids transmission to them. The scheme is analyzed using Markov chains. Numerical results show that, with a correct choice of parameters, the decision taken whether to transmit or not, is correct more than 95% of the time. We have simulated an implementation of the scheme over a wireless system similar to Bluetooth, whose communication model employs frequency hopping over a slotted channel. Results show that the scheme gives significant improvement in goodput even when the threshold parameters are not chosen carefully and the distributions of the good/bad periods are not known. With correct threshold parameters, it yields 40% to 50%

improvement in goodput on the average. It performs better when the wireless environment is more error prone, giving a 150% goodput improvement in the worst case. Test cases which have similar mean durations of the good and the bad periods yield similar results indicating that a good choice of the Reset threshold depends only on the duration of the bad period and not on its distribution.

Future work may involve studying the performance of such a scheme for other wireless systems like 802.11. It would also be interesting to study the behavior of the scheme in conjunction with various scheduling disciplines.

REFERENCES

[1] *The Bluetooth Specification, http://www.bluetooth.org.*

[2] J.C. Haartsen, "The Bluetooth Radio System," *IEEE Personal Communications Magazine*, Feb. 2000.

[3] E.G. Gluck, "Throughput and packet error probability of cellular frequency-hopped spread spectrum radio networks," *IEEE Journal on Selected Areas in Communications*, vol. 7, no. 1, Jan 1989, pp. 148-160.

[4] G. Bark and H.C. Hwang, "Performance comparison of spread spectrum methods on an interference limited FH channel," *IEE Proceedings - Communications*, vol. 146, no. 1, Feb. 1999, pp. 29-34.

[5] T. Ywh-Ren and C. Jin-Fu, "Using frequency hopping spread spectrum technique to combat multipath interference in a multi accessing environment," *IEEE Transactions on Vehicular Technology*, vol. 43, no. 2, May 1994, pp. 211-222.

[6] S.W. Wong and G.F. Gott, "Performance of adaptive frequency hopping modem on an HF link," *IEE Proceedings I - Communications, Speech and Vision*, vol. 137, no.6, Dec. 1990, pp. 371-378.

[7] G. Andersson, "LPI performance of an adaptive frequency hopping system in an HF interference environment," *IEEE 4th International Symposium on Spread Spectrum Techniques and Application Proceedings*, 1996.

[8] M. Inoue, G. Wu and Y. Hase, "Channel state dependent resource scheduling for wireless message transport," *Proceedings of IEEE VTC-Vehicular Technology Conference*, 1998.

[9] M. Inoue, G. Wu and Y. Hase, "Link adaptive resource scheduling for wireless message transport," *Proceedings of IEEE Globecom*, 1998.

[10] M. Inoue, G. Wu and Y. Hase, "Resource scheduling with channel state information for wireless message transport," *IEEE Intl. Conf on Universal Personal Communications*, 1998.

[11] R. Fantacci and M. Scardi, "Performance evaluation of preemptive polling schemes and ARQ techniques for indoor wireless networks," *IEEE Transactions on Vehicular Technology*, vol. 45, no 2, May 1996.

[12] P. Bhagwat, P. Bhattacharya, A. Krishna and S.K. Tripathi, "Enhancing throughput over wireless LANs using channel state dependent packet scheduling," *Proceedings of IEEE Infocom*, 1996.

[13] C. Fragouli, V. Sivaraman and M. Srivastava, "Controlled multimedia wireless link sharing via enhanced class based queuing with channel state dependent queuing," *Proceedings of IEEE Infocom*, 1998.

[14] J.D. Choi, K.M. Wasserman and W.E. Stark, "Effect of channel memory on retransmission protocols for low energy wireless data communications," *IEEE International Conference on Communications*, vol. 3, pp. 1552-1556.

[15] S. Deb, M. Kapoor and A. Sarkar, "Error Avoidance in wireless networks using link state history," *Proceedings of IEEE Infocom*, 2001.

[16] D.K. Anvekar and M. Kapoor, "Frequency lookahead and link-state-history based interference avoidance in wireless pico-cellular networks," *IEEE International Conference on Personal Wireless Communications*, 2000.

[17] J. Zander and J. Malmgren, "Adaptive frequency hopping in HF communications," *IEE Proceedings - Communications*, vol. 40, no. 2, April 1995, pp. 99-105.

[18] J.A. Roberts and J.R. Abeysinghe, "A two state Rician model for predicting indoor wireless communication performance," *IEEE International Conference on Communications*, Seattle WA., 1995.

[19] E.A. Bucher, *UHF satellite communication during scintillation*, Massachusetts Institute of Technology, Cambridge, MA., Tech. Note 175-10, 1975.

[20] R.W. Wolf, *Stochastic Modeling and the Theory of Queues*, Prentice Hall, 1989.

Chapter 6

MAXIMIZING PROFIT IN OPTICAL WDM NETWORKS

Vishal Anand, Chunming Qiao
State University of New York, Buffalo

Abstract: This chapter considers wavelength routed optical networks, which are fast becoming the choice for the backbone of wide-area networks. Such networks employ wavelength routing to establish all-optical data paths called lightpaths. The chapter studies the routing and wavelength assignment of lightpaths with the objective being to maximize the profit. The problem of maximizing profits is formulated as an Integer Programming problem. Heuristic algorithms are also developed to solve the problem assuming both an off-line and an on-line traffic model. Numerical results are presented, which indicate the efficiency of the algorithms.

Keywords: Integer linear programming, Lightpath, Wavelength assignment, Wavelength-conversion, Wavelength routing, WDM networks

1. INTRODUCTION

The recent explosion in Internet traffic and the dawn of the multimedia age implies that the volume of data traffic will far exceed that of voice traffic. This means that the existing telecommunication networks should evolve to match the paradigm shift, from a voice network to data-centric network. The first and most important requirement of future transport networks is the large bandwidth transport capability needed to enable this paradigm shift. In addition, not only is the traffic volume of data communication increasing, but also requirements for new service attributes becoming more tangible. In the future, services that require intensive bandwidth and small delays will become very ordinary. High reliability will also be indispensable for most services.

To develop robust and efficient networks that satisfy the aforementioned requirements, new network architectures and technologies need to be developed. As optical networking technology evolves from megabits per second to gigabits per second to terabits per second and higher, it is needless to say that it is the only way to solve the demands for bandwidth due to exploding Internet traffic volume. The optical layer enables quantum leaps in both transmission capacity and transport node throughput simultaneously by exploiting *wavelength division multiplexing* (WDM) transmission and *wavelength routing*. Optical WDM networks [1, 2, 3] are hence considered to be potential candidates for the next generation of wide-area backbone networks.

In this chapter we study the problem of *routing and wavelength assignment* (RWA) in optical networks. In particular we study the RWA problem with the objective of maximizing the profits.

The chapter is organized as follows. Section 2 defines and explains some of the terms used in optical networks. Section 3 reviews the routing and wavelength assignment problem and also introduces the profitable connection assignment problem. Section 4 and 5 describe heuristics for solving the maximizing profit problem for off-line and on-line traffic respectively. In 4.2, the problem is formulated as an optimization problem, results are presented in 4.3. Section 6 concludes the chapter with a brief summary.

2.   TERMINOLOGY AND CONCEPTS

The *physical topology* of a WDM network consists of routers connected by point-to-point fiber links in an arbitrary mesh as shown in Figure 1. Between two adjacent wavelength-routers, there is a pair of unidirectional fibers (or equivalently, a bi-directional link). Each wavelength-routing node takes in a signal on a wavelength at one of its inputs, and routes it to a wavelength at a particular output, without undergoing opto-electronic conversion. An access station (e.g., an IP router) may be connected to each optical wavelength-router, which can transmit/receive signals through either a tunable transmitter/receiver or a transmitter/receiver array. A connection request is satisfied by establishing a *lightpath* from the source node (access station or wavelength-router) of the connection to the destination node. A lightpath uses one wavelength on each link it spans to provide a circuit-switched interconnection between the source and destination nodes. As shown in Figure 1, two lightpaths, one between nodes 3 and 5, and the other between nodes 3 and 7, must use different wavelengths, (λ_1 and λ_2) on a common fiber link (3 $\rightarrow$ 4), in order to prevent interference of the optical signals.

2.1   Wavelength Conversion

An optical WDM network, in which any lightpath between two nodes should have the same wavelength on all the links its path spans (as shown in Figure 1, is said to have no *wavelength-conversion* capabilities. With wavelength conversion, signals can come in on a wavelength and go out on a different wavelength. However, given current trends and technology, all-optical wavelength-conversion is still an immature and expensive technology, most research assumes that none of the wavelength-routers have wavelength-conversion capabilities. Therefore, a lightpath has to occupy the *same* wavelength on all the links that it spans (this is called the *wavelength-continuity constraint*). Lightpaths set up in such a manner are said to be *all-optical*. In such a wavelength-routed network, the routing and wavelength assignment (RWA) problem consists of two components. The first is a routing scheme that determines the route each of the lightpaths will take. The second component of the RWA problem is to assign a wavelength on each link along the chosen route.

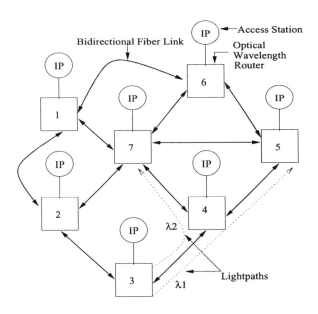

Figure 1. Architecture of a WDM network

Ideally, in a network with M wavelength routers or node, we would like to set up connections between all the $M(M-1)$ pairs. Hence if each node is equipped with $M-1$ transceivers and if there are enough wavelengths on all fiber links, then every node pair can be connected by an all-optical lightpath and there is no RWA problems to solve. However in reality, this is usually not possible because of the limited amount of network resources in terms of available wavelengths and optical switching hardware. Thus only a limited number of lightpaths may be set up in the network.

2.2   Traffic Models

Many different models have been used to describe the traffic demands for connections, and we can classify the connection requests into two categories: *off-line* and *on-line*. In the *off-line* case, we are given the entire set of connections that are to be routed up front. In the *on-line* case, the demands for the connections arrive one at a time, and each connection must be satisfied or rejected without knowing what future demands would be. Typically in communication networks, connection requests arrive according to a *poisson* distribution and the holding times of the connections have an *exponentially* distributed average. A variation of the on-line case is the *incremental* traffic model, wherein connection requests (for a random source and destination) arrive one at a time, and once a connection request is satisfied, the lightpath setup stays in the network for an infinite time.

2.3   Re-routing and Re-arranging of Lightpaths

If we assume that no *re-routing* or *re-arrangement* of lightpaths is allowed in an optical network, then once a route and wavelength is assigned to a lightpath, it is permanent and not allowed to change at a later time. On the other hand if we allow re-routing, the route and wavelength assigned to a lightpath is allowed to change in a dynamic manner so as to accommodate new connection requests. Note that re-routing does not mean that existing connections will be torn down, it only means that the connections will use a different path and/or wavelength.

3.   Routing and Wavelength Assignment

3.1   Survey of Related Work

In the past many RWA heuristics [1, 2, 4, 5, 6, 7, 8, 9] have been proposed to optimize different criteria. For instance, [1, 2] develop and survey several practical approaches to solving the RWA problem. [4] focused on the design of a logical topology and the routing algorithm so as to maximize the network throughput while constraining the average delay seen by source-destination nodes and the amount of processing required at the nodes. [5] formulated the RWA problem as an optimization problem with the following objective functions: 1) given a traffic matrix, minimize the network-wide average packet delay; 2) maximize the scale factor by which the traffic matrix can be scaled up to provide maximum capacity upgrade for future traffic demands. [6] considered the objective of minimizing the average weighted propagation delay. In [7] the objective was to maximize the amount of traffic carried in one hop from the source to the destination, but with degree and delay constraints. More recently, [8] investigated the problem of routing connections in all-optical networks while considering degradation of routed signals by different optical components. [9] presented a global methodology and discussed several optimization options for dimensioning the WDM layer.

[10, 11] consider *fixed routing* with which every source-destination pair is associated with a single route, and thus a connection request is *blocked* if no resources (e.g., wavelengths) are available along the associated route. *Alternate routing* schemes have also been explored in [12], which approximate the behavior of dynamic routing algorithms [13]. Wavelength assignment schemes such as *fixed-order, first-fit, random-order, wavelength-packing* have also been researched in [13, 14, 15]. In [16], we studied the dynamic establishment of protection paths by allowing for re-arrangement in survivable networks.

3.2 The Profitable Connection Assignment Problem

In the past, many variants of the RWA problem have been studied with various optimization criteria (e.g. minimizing traffic delay, maximizing network throughput). However, what is more important for a *carrier or bandwidth broker* is the *profit* that they manage to make. Although minimizing costs or maximizing throughput (connections) do indirectly serve this goal, it can be argued that a case exists for a more direct study of the problem of maximizing profits.

The problem of maximizing profit is modeled as follows:

3.3 Problem Formulation

Given:

1 A physical network topology modeled as a weighted undirected graph $G = (V, A)$, where V is the set of network nodes and A is the set of links connecting these nodes. The weight on each link is the cost incurred while using a wavelength on that particular link.

2 The number of wavelengths available on each link, λ.

3 An NXN traffic matrix T, where $N = |V|$, number of nodes in the network. The $T_{i,j}$ indicates the number of connection (or lightpaths) requested between source node i and destination node j. This traffic matrix can be asymmetric, i.e. $T_{i,j} \neq T_{j,i}$.

4 An earnings matrix E, where the $E_{i,j}$ indicates the earnings obtained by satisfying one connection request between source i and destination j. This matrix is also asymmetric.

The total profit that can be made is defined as *P=(Total Earnings - Total Cost)*. Each connection request i, $1 \leq i \leq M$ (*M* being the total number of connection requests), is associated with a fixed earning E_i and incurs a varying cost C_i, which depends on the route taken from the source to the destination. The objective is to satisfy $n \leq M$ connections such that P is maximized. Note that when $n = M$, i.e., if all the connections between end nodes can be satisfied, the above reduces to the *minimizing cost* problem. If $E_i = E$, and $C_i = C$, then it becomes the *maximizing throughput* problem. However, we will consider the more practical case where the cost of a connection depends on its route, $E_i \neq E_z (i \neq z)$ and $n \leq M$.

3.4   Network Model

The network topology considered in this chapter is the NSFNET backbone shown in Figure 2. We assume that the cost of using any wavelength on a particular link is the same, but may vary from link to link. Let c_i be the cost of using each wavelength on link i, then $C_{\alpha\beta} = \sum_i c_i$ is the total cost (summation of the cost of using each link along a path) of establishing a connection between nodes α and β. We assume no wavelength-conversion capabilities at any of the nodes and single hop communication. Hence to meet a connection request, the routing and wavelength assignment must be done such that the lightpath set up to satisfy the connection request uses the same wavelength on all the links it spans. We also assume that the number of wavelengths on each link in the network is the same. Note that these assumptions do not affect the applicability of the proposed heuristics.

4.   OFF-LINE APPROACH

It is easy to show that the optimal routing problem is NP-complete by using the results of [10] and by restricting the general problem to tree topologies. Further the problem of wavelength assignment to lightpaths can be reduced to the *vertex coloring* problem in a graph, which is also NP-complete [17]. Hence heuristic based approaches are often used to solve the problem.

4.1   Heuristic Solution for Off-line Traffic

4.1.1   Heuristic 1: *Minimizing cost heuristic.*

1 Find the cheapest path using the Dijkstra's algorithm [18] between each node pair, which has at least one connection request in the traffic matrix.

2 Sort the above paths in the order of *increasing* cost and store them in a list, Z.

3 Select the cheapest path from Z (i.e., the first on the list), say from i to j. If a wavelength is found such that it is unused on all the links the path spans, assign that wavelength to establish a lightpath/connection

from i to j, and delete the path from the list. If not, find a cheapest path from i to j currently available taking into consideration that some of the wavelengths may have been used to satisfy other connections. Insert this new path into the sorted list Z depending on its cost. Repeat this step till no other connection requests can be satisfied. If no wavelength can be assigned to any of the paths, the connection request is dropped.

4 Compute the overall profit $P_1 = B - C$, where B is the earnings got from satisfying the above set of connections and C is the cost incurred for the same.

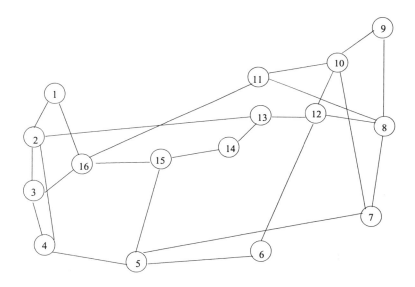

Figure 2. 16 node NSFNET topology

4.1.2 Heuristic 2: *Maximizing profit heuristic - MaxPro.*

1 Find the cheapest path using the Dijkstra's algorithm between each node pair, which has at least one connection request in the traffic matrix.

2 Sort the above paths in *descending* order of profit (earnings – cost) and store them in a list, Z.

3 Select the maximum profit path from Z, say from i to j. If a wavelength is found such that it is unused on all the links the path spans, assign that wavelength to establish a connection and delete the path from the list. If not, find a cheapest path from i to j currently available. Insert this

new path into the sorted list Z. Repeat this step till no other connection requests can be satisfied.

4 Compute the overall profit $P_2 = B - C$.

4.1.3   Heuristic 3: *Random Assignment heuristic.*

1 Find the cheapest path using the Dijkstra's algorithm between each node pair, which has at least one connection request in the traffic matrix.

2 Choose a node pair (i, j) at random, from among those which have at least one connection request to be satisfied. Try and assign a wavelength to establish a lightpath from i to j, as before. Repeat this step till no other connection requests can be satisfied.

3 Compute the overall profit $P_3 = B - C$.

4.2   Optimization Problem

We formulate and solve the problem as an Integer Linear Program (ILP), using the following notation:

1 α and β used as subscript denote the *source* and *destination* of a connection request respectively.

2 r denotes a particular (r^{th})route number between a source-destination pair.

3 k denotes the kth wavelength.

Given:

- $G(V, A)$, an undirected graph, where V is the set of nodes, A is the set of links.

- $n = |V|$, $L = |A|$, n and L are the number of nodes and edges in the graph respectively.

- D, the traffic matrix, in units of flow; $d_{\alpha\beta} = 2$,, implies that two connection/lightpath requests exists between nodes α and β.

- C, the cost matrix; $c_{\alpha\beta} = \infty$, if edge $(\alpha, \beta) \notin A$, else $c_{\alpha\beta}$ is finite.

- E, the earnings matrix; $e_{\alpha\beta}$ indicates the earnings obtained by satisfying one connection request between nodes α and β.

- λ, the number of wavelengths on each edge of G.

Constants:

- $R_{\alpha\beta}$, the total number of alternate routes (paths) available to reach node β from node α.

- $C_{\alpha\beta}^r$, the cost incurred to reach node β from node α on route number r.

- $l_{\alpha\beta}^{rj}$, $l_{\alpha\beta}^{rj} = 1$, if link j is used by route number r between nodes α and β and 0 otherwise.

Variables:

- $x_{\alpha\beta}^{rk}$: $x_{\alpha\beta}^{rk} = 1$, indicates that the connection request between nodes α and β can be satisfied by setting up a lightpath routed on route r, using wavelength k, $x_{\alpha\beta}^{rk} = 0$ indicates that the kth wavelength is not available for this route.

Constraints:

-

$$\sum_{r=1}^{R_{\alpha\beta}} \sum_{k=1}^{\lambda} x_{\alpha\beta}^{rk} \leq d_{\alpha\beta} \qquad \forall \alpha, \beta$$

The total number of lightpaths established between a node pair (α, β) should not exceed the number connection requests, $d_{\alpha\beta}$ in the traffic matrix D.

-

$$\sum_{\alpha=1}^{n} \sum_{\beta=1}^{n} \sum_{r=1}^{R_{\alpha\beta}} l_{\alpha\beta}^{rj} \sum_{k=1}^{\lambda} x_{\alpha\beta}^{rk} \leq \lambda \qquad \forall j = 1, ..., L$$

For every link $j \in A$, the number of lightpaths established on j should not exceed the wavelength capacity, λ of the link.

-

$$\sum_{\alpha=1}^{n} \sum_{\beta=1}^{n} \sum_{r=1}^{R_{\alpha\beta}} l_{\alpha\beta}^{rj} x_{\alpha\beta}^{rk} \leq 1 \qquad \forall j = 1, ..., L, \forall k = 1, ..., \lambda$$

Any wavelength k on any link j, can support at most one lightpath.

-

$$x_{\alpha\beta}^{rk} = (0, 1) \qquad \forall \alpha, \beta, r, k$$

The integrality constraint.

Objective: Optimality Criterion

-

$$Maximize : \quad \sum_{\alpha=1}^{n} \sum_{\beta=1}^{n} \left[E_{\alpha\beta} \sum_{r=1}^{R_{\alpha\beta}} \sum_{k=1}^{\lambda} x_{\alpha\beta}^{rk} - \sum_{r=1}^{R_{\alpha\beta}} C_{\alpha\beta}^{r} \sum_{k=1}^{\lambda} x_{\alpha\beta}^{rk} \right]$$

The above objective function maximizes the total overall profit $P=$ *Total Earnings - Total Costs* by solving the RWA problem. This objective function may also be applied to the on-line case (to be discussed later), assuming that re-arrangement of lightpaths is allowed.

4.3   Numerical Results

4.3.1   Comparison of the heuristics.

The above described 3 heuristics were used to derive results for the maximizing profit problem. All simulations were done on the NSFNET backbone, the number of wavelengths on a link varied from 1 to 16. The earnings associated with a connection request and the costs associated with each link varied according to a normal distribution. The number of connection requests between node pairs varied according to a uniform distribution. The overall profit obtained from each of the three heuristics for different earnings and costs was then compared. Some of our results are tabulated in Table 1. Results indicate that MaxPro almost always performs better than Heuristics 1 and 3, when the objective is to maximize the profit. The difference in the performance of Max-Pro and Heuristic 1 is due to the fact that at each step MaxPro tries to maximize the profit by taking the earnings into consideration while also minimizing the cost, but Heuristic 1 always tries to minimizes the cost without taking the earnings into consideration, which does not always maximize the profit. Heuristic 3 is used for comparison purposes and we observe that MaxPro and Heuristic 1 always perform better.

Table 1.   Summary of experimental results, comparison of Heuristic 1, MaxPro and Heuristic 3

No. of Wavelengths/link	Heuristic 1.	MaxPro	Heuristic 3.
1	389	440	197
4	1006	1183	615
8	1539	1762	1058
10	1712	1956	1207
12	1885	2128	1388

4.3.2   Comparison of MaxPro and ILP.

The optimization problem was solved using Integer Linear Programming using LINDO [19] on a Sun Ultra-60. A sample network topology that we considered is shown in Figure 3 (the reason we chose this network instead of the NSFNET is due to the limit on the number of variables and constraints imposed by LINDO). We assumed a uniform distribution for the connection requests and a normal distribution for the costs and earnings. Simulations with $\lambda = 2$ and $\lambda = 4$ were conducted, the number of connection requests between node pairs was varied from 1 to 8 for each value λ, the results are tabulated in Table 4 and 5. A sample cost matrix is shown in Table 2, an entry in the table denotes the cost of using the corresponding link in the network. If a link does not exist the entry is ∞. Similarly a sample earnings matrix is shown in Table 3.

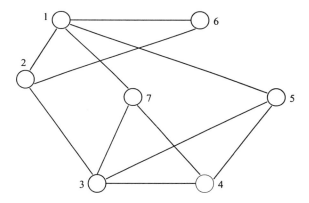

Figure 3. Network topology for the comparison of Maxpro and ILP

Table 2. Sample link cost matrix

Node Number	1	2	3	4	5	6	7
1	∞	12	∞	7	∞	15	11
2	15	∞	12	∞	7	∞	13
3	∞	9	∞	4	3	5	∞
4	∞	3	15	∞	∞	11	∞
5	14	∞	7	∞	∞	11	∞
6	14	∞	7	∞	13	∞	∞
7	12	13	∞	11	∞	∞	∞

The first column in Table 4 and 5 shows the number of requested connections for each node pair. The next two columns give the maximum profit obtained for $\lambda = 2$ by the ILP and the heuristic, MaxPro respectively. The fourth column compares MaxPro (M) and ILP as a ratio. The last three columns give the corresponding values of maximum profit for $\lambda = 4$. From these results, we

Table 3. Sample earnings matrix

Node Number	1	2	3	4	5	6	7
1	0	22	40	55	38	32	55
2	28	0	34	46	25	41	41
3	56	44	0	23	30	51	24
4	55	47	27	0	35	56	50
5	49	31	33	52	0	22	20
6	58	53	51	50	50	0	42
7	28	39	56	20	27	32	0

Table 4. Comparison of MaxPro and ILP, $\lambda = 2$

No. of requests/node pair	ILP $\lambda = 2$	MaxPro(M) $\lambda = 2$	M/ILP(%)
1	720	693	96%
2	844	774	91%
3	844	775	91%
4	844	776	91%
6	844	776	91%
8	844	776	91%

Table 5. Comparison of MaxPro and ILP, $\lambda = 4$

No. of requests/node pair	ILP $\lambda = 4$	MaxPro(M) $\lambda = 4$	M/ILP(%)
1	1052	944	90%
2	1440	1386	96%
3	1564	1421	91%
4	1688	1548	92%
6	1688	1548	92%
8	1688	1552	92%

observe that the heuristic approach gives a profit that is, on the average, at least 90% of the optimal maximum value obtained from the ILP and in many cases more than 90%. The maximum number of connections that can be satisfied between any node pair (i, j) in the network is limited by the number of wavelengths going out of source i and coming into the destination j. This explains the fact that if the number of wavelengths per link are doubled (from 2 to 4), the profits almost double due to the fact that more connections can now be satisfied.

The difference in the performance of MaxPro and ILP is due to the fact that the heuristic is greedy and at each step only tries to satisfy a connection request which will give the maximum profit. Thus it is shortsighted and may satisfy a connection, which later on blocks other connections, which if had been satisfied instead would have given a higher overall profit. Thus, the heuristic in the process of maximizing the profit at each step may not achieve the maximum possible overall profit. ILP on the other hand avoids this problem by trying all possible combinations and then chooses those connections which would lead to the maximal overall profit. However, ILP becomes computationally intractable for even moderately large networks and hence we have to resort to heuristics.

5. ON-LINE APPROACH

As discussed earlier, in the off-line case all the connection requests are known prior to making any routing and wavelength assignment decision. Hence, an appropriate order and combination of some connection requests can be chosen so as to maximize the profit [20]. In the on-line case the connection requests arrive one at a time and a decision as to whether a request can be satisfied, and if so how, has to be made without the knowledge of future connection requests. As a result, we expect that the profit obtained in the on-line case is much lower than that obtained from the off-line case, which implies that there may be room for improvement in the on-line case. In [21] and the following sections we discuss such approaches. In Table 6 we tabulate a sample of the

Table 6. Comparison of the off-line and on-line heuristics

No. Wavelengths/link	Off-line	On-line	On-line/Off-line(%)
1	3400	1230	36%
4	12650	7370	58%
8	25400	15090	60%
12	44360	21170	48%
16	55500	29380	53%

results obtained by off-line and on-line heuristics. In the off-line case we used a heuristic, MaxPro that sorts all requested connections in the decreasing order of potential profits, and then tries to satisfy the most profitable connection first [20]. In the on-line case, we used a simple heuristic that assigns a newly arrived connection on its cheapest available path and drops the connection request if no path is available without attempting to re-route any existing connections. As expected the on-line heuristic yields a much lower profit than the off-line heuristic.

In an attempt to improve the performance of the on-line case, in the next section, we study the performance of heuristics, which re-route some of the existing connections.

5.1   Re-routing Heuristics

In this section, we present on-line heuristics, which allow re-routing of existing connections when a new connection request is blocked to improve the profits. For each of these heuristics, we assume that once a connection request has been satisfied it stays in the network for an infinite time. In other words, re-routing does not mean that existing connections will be torn down, it only means that the connections will use a different path and/or wavelength. In addition, re-routing of existing connections is done *only* when no lightpaths can be found in the network for the new connection request.

It helps to think and model the network consisting of λ wavelengths on each link as λ identical graphs one for each wavelength. Hence the network can now be thought of as λ identical *wavelength-graphs* [22], each representing a wavelength and each physical link has a corresponding link in each of the wavelength graphs. The following four heuristics are considered.

5.1.1   Heuristic 1:Shortest-path Min-Disruptions (SPMD).

1 Find the shortest (cheapest cost) path for the given connection request using Dijkstra's algorithm in the graph G.

2 In each of the wavelength graphs, find the number of already satisfied connections, which will be disrupted if this request is assigned that wavelength.

3 Pick the graph with the minimum disrupted connections. Temporarily satisfy this new connection request in that graph (i.e. on that wavelength).

4 Sort the disrupted connections in the descending order of their earnings. Try to satisfy each (temporarily) using a path that is cheapest at that instant.

5 If we can successfully re-route all of the disrupted connections and the resultant profit is greater than the profit made prior to satisfying this new connection. Make the above assignments and re-routing.

6 If we cannot re-route all of the disrupted connections or the resultant profit is lower due to re-routing, reject this new connection request.

The probability that a new connection request will be satisfied is high, if we assign it a path and wavelength that causes minimum number of disruptions. The heuristic is greedy, as it always tries to assign the new connection on the cheapest and hence most profitable path. On the other hand, it may be short-sighted as the re-routed connections may have to take longer and less profitable paths finally, hence causing the overall profit to be sub-optimal.

5.1.2 Heuristic 2:k Shortest-paths Min-Disruptions (k-SPMD).

For each new connection request we find k, $k > 1$ disjoint shortestpaths between the respective node pairs. This is similar to heuristic 1, but instead of limiting to only one shortestpath (step 1 in Heuristic 1), we try k disjoint shortestpaths. The shortestpath that causes minimum disruptions is then used to satisfy the new connection request.

There could be alternate paths for a particular connection request that are more expensive but causes lesser disruptions of older connections. This heuristic tries to improve over the first one, by trying to be less greedy and assigns the new connection on that path which is not necessarily the cheapest, but causes the least number of disruptions. Hence improving the probability that the new connection request is satisfied and the overall profit is increased.

5.1.3 Heuristic 3:Shortest-path Max-Profit (SPMP).

This is similar to heuristic 1, but instead of limiting only to the wavelength graph that causes minimum disruptions, we try and assign the connection request on its shortestpath on every wavelength graph irrespective of the number of disruptions caused and then finally choose that one which results in maximum profit.

It is possible that the profit on the minimum disruptions wavelength graph is lower than that which could have been obtained by assigning the new connection on a wavelength that causes more disruptions. This is because the re-routed connections are different in the two cases and although the number of connections that have to be re-routed are more in the latter case they take cheaper paths. This heuristic takes this fact into account and assigns the new connection on that wavelength graph which provides maximum profit. This heuristic also increases the probability that a new connection request is satisfied as it tries to assign it on all the wavelength graphs.

5.1.4 Heuristic 4:Shortest-path Max-Profit d-Disruptions (SPMP-d).

This is a variation of heuristic 1 and subset of heuristic 3. The difference with heuristic 1 being that instead of picking the wavelength graph with minimum disruptions as in step 3, we try to assign the new connection on all those wavelengths which cause less than d number of disruptions of previously satisfied connections. The profit is computed for each case and finally that wavelength graph is chosen which gives the maximum profit.

The heuristic is less exhaustive than heuristic 3. It tries to be less greedy than heuristic 1 and takes into account the fact that minimum number of disruptions does not always lead to maximum profit.

5.2   Results and Comparison

In our study we simulated and compared the results of the four heuristics, SPMD, k-SPMD, SPMP and SPMP-d. The results are tabulated in Table 7, in the Table, P_T is the total profit obtained, C_s is the number of connections satisfied and C_r is the number of connections satisfied via re-routing.

Table 7.   Comparison of the 4 on-line heuristics

No.	SPMD			k-SPMD($k>1$)			SPMP			SPMP-d($d=3$)		
of λ	P_T	C_s	C_r	P_T	C_s	C_r	P_T	C_s	C_r	P_T	C_s	C_r
1	1230	10	0	1230	10	0	1280	12	4	1260	11	2
4	6920	50	1	7370	51	0	7450	54	3	7270	51	2
8	14620	87	1	14680	106	2	15020	116	6	14860	109	4
12	20940	151	0	21450	152	1	21960	163	6	21620	158	5
16	28180	203	0	28700	204	1	29360	215	5	29360	215	5

The results obtained can be explained as follows. It is easy to see that SPMD is a subset of k-SPMD and SPMP-d is a subset of SPMP. Also SPMD is a subset of SPMP-d. This is because we are restricting the domain of search as we go from k-SPMD to SPMD and SPMP to SPMP-d. The results obtained reflect this. We observe that the profit obtained and connections satisfied by a superset heuristic is always more than its subset. The trade-off being an increase in computational complexity and complex logic of the superset heuristic. Also, it can be seen from Table 7 that although the heuristic SPMP gives the best results, the difference is very small when compared to the other simpler heuristics.

Table 8.   Comparison of SPMP with on-line non re-routing heuristic

No. of λ	On-line non Re-routing		SPMP	
	P_T	C_s	P_T	C_s
1	1230	10	1280	12
4	7370	51	7450	54
8	15090	106	16020	116
12	21170	152	21960	163
16	29380	205	31360	215

We then compare SPMP with the on-line heuristic without re-routing as described in Section 5. The results are shown in Table 8. The resulting profit and number of connections requests satisfied in each case are observed to be similar. This is due to the fact that the number of re-routing occurring (as a percentage of the number of connections requests satisfied) is very small. These results can be explained as follows. Each of our re-routing heuristics, initiate re-routing *only* when no path on any of the wavelengths can be found for a connection request. Thus by the time a connection request is blocked and we try to satisfy this request by re-routing some of the existing connections, the network load is already high. So in most of the cases it is very difficult to find an alternate path by re-routing for the disrupted set of connections, in which case the new connection request is dropped. Further, it cannot be guaranteed that the re-routing is beneficial to the future connection requests and overall profit. Some results (e.g. $\lambda = 12, 16$) in the above table show that, although by re-routing (using SPMP) the number of connection requests satisfied increases, it does not increase overall profit significantly. The results hence indicate that re-routing of existing connections in a highly loaded network only when a connection request is blocked does not help significantly in increasing the overall profit.

6. SUMMARY

Optical networks using wavelength division multiplexing (WDM) technology are fast becoming the choice for the next generation of the Internet backbone. This chapter studied the routing and wavelength assignment (RWA) problem as applied to a wavelength-routed, all-optical networks. The objective was to maximize the profit that could be made by a carrier or bandwidth broker, given a certain set of lightpath requests that were to be satisfied on a given physical topology. The problem has been considered for two traffic models, off-line and on-line (with incremental traffic). It was shown that the problem of maximizing the profit is different from the traditional problem of minimizing costs or maximizing throughputs. Heuristic algorithms have been proposed to solve the maximizing profit problem. The problem has also been formulated as an Integer linear program, to obtain the optimal solution under the off-line traffic model.

We have also compared heuristics for on-line (with incremental traffic model) traffic which do not allow for re-routing or re-arrangement of lightpaths with the off-line algorithms. The results indicate that in the off-line case one can achieve a much higher profit than in the on-line case. Then various on-line heuristic algorithms, which allow for re-routing of existing connections have been presented. However, the performance of these re-routing algorithms when compared to the on-line algorithms, which do not do any re-routing was similar. This behavior is attributed to the fact that re-routing of existing connections is done only when blocking of new requests occurs.

REFERENCES

[1] B. Mukherjee, *Optical Communication Networks*, New York, NY: McGraw-Hill, 1997.

[2] R. Ramaswami and K. N. Sivarajan, *Optical Networks: A practical Perspective*, San Francisco, CA: Morgan Kaufmann, 1998.

[3] P. E. Green, *Fiber Optic Networks*, Englewood Cliffs, NJ: Prentice Hall, 1993.

[4] R. Ramaswami and K. N. Sivarajan, "Design of Logical Topologies for Wavelength-Routed All-Optical Networks," *IEEE/ACM transactions on networking*, 1995.

[5] B. Mukherjee, S. Ramamurthy, D. Banerjee and A. Mukherjee, "Some Principles for Designing a Wide-Area Optical Network," in *Proc., IEEE INFOCOM*, 1994.

[6] I. Chlamtac, A. Ganz and G. Karmi, "Lightnets: Topologies for High-speed Optical Networks," *IEEE/OSA J. of Lightwave Technology.*, 1993.

[7] Z. Zhang and A. S. Acampora, "A Heuristic Wavelength Assignment Algorithm for Multihop WDM Networks with Wavelength Routing and Wavelength Reuse," in *Proc., IEEE INFOCOM*, 1994.

[8] M. Ali, B. Ramamurthy and J. S. Deogun, "Routing Algorithms for All-Optical Networks with Power Considerations: The Unicast Case," in *Proc., IEEE International Communications and Networks*, 1999.

[9] M. Garnot, M. Masetti, F. Nederlof, L. Eilenberger, G. Bunse and S. Aguilar, "Dimensioning and Optimization of the Wavelength-Division-Multiplexed Optical layer of Future Transport Networks," in *Proc., IEEE International Conference on Communications*, 1998.

[10] I. Chlamtac, A. Ganz and G. Karmia, "Lightpath Communications: An Approach to High Bandwidth Optical WAN's," *IEEE Transactions on Communications*, 1992.

[11] A. Girard, *Routing and Dimensioning in Circuit-Switched Networks*. MA: Addison-Wesley, 1990.

[12] V. Anand and C. Qiao, "Static versus Dynamic Establishment of Protection Paths in WDM Networks," *J. of High Speed Networks (Special Issue on Survivable Optical Networks* Vol.10, No.4, 2001, pp.317-327.

[13] T. Wu, *Fiber Network Service Survivability: A Comprehensive Approach*. Norwood, MA: Artech House, 1992.

[14] T. Wu, "Emerging Technologies for Fiber Network Survivability," in *IEEE Comm. Mag.*, vol. 33, pp. 58–74, February 1995.

[15] S. Ramamurthy and B. Mukherjee, "Survivable WDM Mesh Networks, Part I-Protection," in *Proc. IEEE INFOCOM*, 1999.

[16] V. Anand and C. Qiao, "Dynamic Establishment of Protection Paths in WDM Networks, Part I," in *IEEE ICCCN 2000 (The 9th International Conference on Computer Communications)*, Las Vegas, Nevada, 2000.

[17] M. R. Garey and D. S. Johnson, *Computers and Intractability: A Guide to the Theory of NP-Completeness*, 1979.

[18] A. V. Aho, J. E. Hopcroft and J. D. Ullman, *Data Structures and Algorithms*, Addison-Wesley, 1987.

[19] http://www.lindo.com.

[20] V. Anand, T. Katarki and C. Qiao, "Profitable Connection Assignment in All-Optical WDM Networks," in *IEEE/ACM Workshop on Optical Networks, Dallas, TX*, 2000.

[21] V. Anand, T. Katarki and C. Qiao, "Profitable Connection Assignment for Incremental traffic in All-Optical WDM Networks," in *Academia-Industry Working-conference on Research Challenges (AIWoRC), Buffalo, NY*, 2000.

[22] L. H. Sahasrabuddhe and B. Mukherjee, "Light-Trees: Optical Multicasting for Improved Performance in Wavelength-Routed Networks," in *IEEE Comm. Mag.*, vol. 37, pp. 67–73, February 1999.

Chapter 7

A TAXONOMY OF DATA MANAGEMENTS VIA BROADCASTING IN MOBILE COMPUTING*

Ilyoung Chung, Bharat Bhargava, Sanjay K. Madria
Dept. of Computer Sciences, Purdue University; Dept. of Computer Sciences, Purdue University; Dept. of Computer Science, University of Missouri-Rolla

Abstract: Data management for distributed computing has spawned a variety of research work and commercial products. At the same time, recent technical advances in the development of portable computing devices and the rapidly expanding cordless technologies have made the mobile computing a reality. In conjunction with the existing computing infrastructure, data management for mobile computing gives rise to significant challenges and performance opportunities. Most mobile technologies physically support broadcast to all mobile users inside a cell. In mobile client-server models, a server can take advantage of this characteristics to broadcast information to all mobile clients in its cell. This fact introduces new mechanisms of data management which are different from the traditional algorithms proposed for distributed database systems. In this chapter, we give executive summary and discuss topics such as data dissemination techniques, transaction models and caching strategies that utilize broadcasting medium for data management. There is a wide range of options for the design of model and algorithms for mobile client-server database systems. We present taxonomies that categorize algorithms proposed under each topic. Those taxonomies provide insights into the tradeoffs inherent in each field of data management in mobile computing environments.

Keywords: Broadcasting, Cache Invalidation, Data Dissemination, Transaction management

*This research is supported by CERIAS and NSF grants CCR-9901712 and CCR-0001788

1.   INTRODUCTION

The widespread adoption of a client-server system has made this architecture the conventional mode for distributed database systems. At the same time, recent technological advances in the development of portable computing devices and rapidly expanding cordless technologies have made the mobile computing a reality. The combination of these factors gives rise to significant challenges and performance opportunities in the design of mobile client-server database systems. Broadcasting is widely accepted as a medium of disseminating information from the server to mobile clients in a cell. In this chapter, we consider three important topics which utilize broadcasting facility in mobile computing environments: the management of transactions which access data using mobile devices, data dissemination which adopts broadcast as a means of delivering data to mobile devices, and data caching strategies which lessen the dependency on the server.

Approaches to handling weak connectivity in data management systems aim at minimizing communication and surviving short disconnections. However, due to the complicated dependencies among database items, the problem is a complex one. Algorithms on such topics have been proposed to achieve high performance while surviving resource restrictions of resources and connections using broadcasting facility. In this chapter, we present taxonomies that describe the design spaces for transaction management, data dissemination, and caching for data management, and show how proposed algorithms are related to one another. We provide a unified treatment of proposed algorithms that can be used to help understanding the design alternatives and performance tradeoffs for each topic.

The rest of this chapter is organized as follows. In Section 2, alternative architectures for mobile data management are discussed. In Section 3, the taxonomy of concurrency control algorithms which utilize broadcasting is presented and tradeoffs for each level of design options are discussed. In Section 4, we shall focus on mechanisms for data dissemination which broadcasts data to all mobile devices inside a cell. In Section 5, taxonomy for cache invalidation schemes is proposed and discussed, and finally we state our concluding remarks in Section 6.

2.   ARCHITECTURAL ALTERNATIVES

A mobile client-server database architecture can be categorized according to some design alternatives, which will be discussed in this section. A mobile client-server database architecture can be categorized according to the four design alternatives namely 1) *the unit of interaction between clients and server*, 2) *data delivery method*, 3) *caching at mobile client*, and 4) *communication method.* The tradeoffs of these architectural alternatives will be discussed in this section.

2.1 The Unit of Interaction

A mobile client-server database architecture can be categorized according to the unit of interaction between mobile clients and servers. In general, mobile clients can send data requests to the server as *queries* or as *requests for specific data item*. Systems of the former type are referred to as *query-shipping* system and those of the latter type are referred to as *data-shipping*. In query-shipping systems, a client sends a query to the server. The server then processes the query and sends the results back to the client. Queries may be sent as plain text (e.g.,SQL), in a compiled representation, or as calls to precompiled queries that are stored at the server. In query-shipping architecture, communication costs and client buffer space requirements are reduced since only the data items that satisfy a given query are transferred from the server to clients. In contrast, data-shipping system perform the bulk of the work of query processing at the clients, and as a result, much more database functionality is placed at the clients. It offloads functionality from the server to mobile clients. This may prove crucial for performance, as the computing power and the amount of resources at mobile clients get stronger and increase. Also, the frequency of communications between servers and mobile clients reduces, as the applications at mobile clients access data items which resides in the mobile clients. Reduced frequency of communications is crucial for the performance of mobile DBMSs, considering the limited bandwidth of wireless channel. These two alternatives are shown in Fig. 1.

2.2 Data Delivery Method

In a data-shipping client-server system, the server is the repository of the data and the clients are the consumers of the data. Thus, when a client application requires a data item, it has to be delivered from the server to the client. Broadly, there are two ways to achieve the data transfer.

In the client-initiated approach, the client requests a data item from the server on demand, i.e., when the client application requests it. In response to the request, the server locates the data item and transfers it to the client. To make this transfer feasible, the clients and the server typically use a mutually agreed upon request/response protocol. As part of the protocol, the client is allowed to make a predetermined set of request to which the server responds appropriately. On the other hand, in the server-initiated approach, the responsibility of transferring the data rests with the server. Here, the server delivers data items to mobile clients without any explicit request from it, and in anticipation of an access in the future. Thus, unlike in the first approach, the transfer

Query Shipping

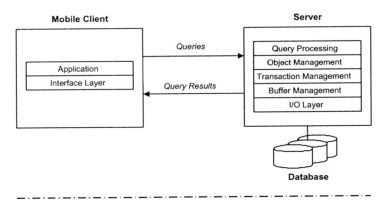

Data Shipping

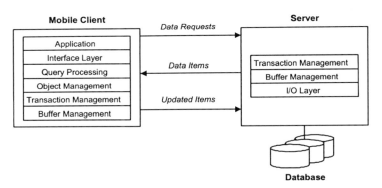

Figure 1. Architectural alternative: Unit of interaction

is initiated by the server and not by the client. Thus the server has to have some knowledge of the client data requirements in this case.

The client-initiated delivery is also called *pull-based delivery* and systems based on it are termed pull-based systems. In effect, in this approach, clients *"pull"* data from the server on demand. Conversely, the data delivery using the server-centric approach is called *push-based delivery*, since the server *"pushes"* data out to clients. These two different alternatives are shown in Fig. 2.

2.3   Caching at Mobile Clients

Client caching refers to the ability of mobile clients to retain copies of data items locally once they have been obtained from the server. In client-server

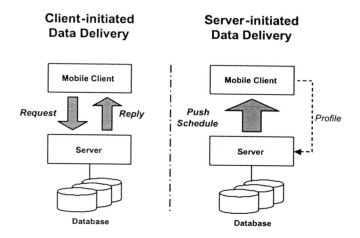

Figure 2. Architectural alternative: Data delivery method

database systems, client caching can be characterized by the following concepts: *dynamic replication* and *second class ownership*. Dynamic replication means that data copies are created and destroyed based on the runtime demands of clients. When a client needs to access a data item, a copy of that item is placed in the mobile client's cache if it does not already exist. In some schemes, data items are removed from a client's cache in order to make room for more recently requested ones because they become invalid. Therefore, the amount of replication that exists in the system at any given time is driven by the recent activity of the mobile clients. This is in contrast to static replication in which the replication of data is determined as part of the physical database design process.

It is well known that replication can reduce data availability in the presence of updates and failures in a distributed environments. Second-class ownership allows consistency to be preserved without sacrificing availability. Second-class ownership refers to the fact that in client caching, the cached copies of pages are not considered to be the equals of actual data items, which are kept at the server.

Mobile client caching is a compromise between the utilization of mobile client resources and the correctness and availability concerns. Mobile client resources are exploited by maintaining the cache, thereby reducing the need to obtain data from the servers. Data that is in a mobile client's local cache can typically be accessed faster than data that is at a server, considering the limited bandwidth of wireless links. Caching, however, does not transfer ownership of data to mobile clients. Servers remain the only true owners of data, and therefore are ultimately responsible for ensuring the correctness of transaction execution. Client autonomy is sacrificed in the sense that servers maintain data ownership and must remain involved in the execution of transactions. However, the pay off here is that the server's participation in transactions execution is minimized.

2.4   Communication Method

In mobile client-server database systems, the server should maintain communication to mobile clients in order to transfer many control information as well as data items. Those control information are used at mobile clients in processing transactions, which include control of concurrent transactions and maintenance of data consistency. First, the server should send concurrency control information that includes conflict relations of transactions which are executed concurrently at mobile clients. Also, the server should notify a transaction about the commit result of the certification process, according to the correctness criteria. Secondly, update information of data items should be transferred to mobile clients in order to ensure the consistency between the server and mobile clients. When mobile clients maintain local cache and keep a portion of data, applications at mobile clients assume that the local copies of data have up-to-data value of those items. In order to satisfy such assumption, consistency information should be sent to mobile clients, whenever data items are updated.

For sending the control information from the server to mobile clients, there are two communication methods; *unicast* and *broadcast*. With unicast method, the server sends each control information to a specific mobile client, with the knowledge about the clients' information. This method requires that a mobile client register its presence and that a server keep information about mobile clients. The server in this case is *stateful* since it knows about the state of the mobile clients. On the other hand, with broadcast method, the server sends the control information to all the mobile clients periodically or aperiodically, with the same message. Since clients may require different control information simultaneously, the broadcasting message should be well defined. The server in this case is *stateless* since it does not know about the state of mobile clients.

3.   TRANSACTION MANAGEMENT

In mobile client-server database systems, transactions are initiated by each mobile client as a string of read and write operations. Because multiple transactions which access common data items may be issued concurrently, there should be a protocol which guarantees the correct execution of concurrent transactions. Several protocols have been proposed in the literature to control concurrent transactions in mobile database systems, and those protocols

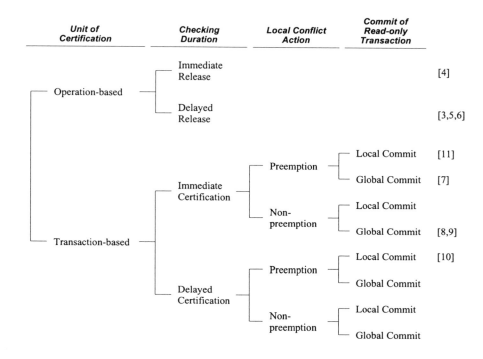

Figure 3. Taxonomy of concurrency control schemes

can be evaluated based on some criteria. In this section, considering the characteristics of mobile environments, we define the following:

- **Concurrency:** The degree of concurrency represents the ability of a protocol which can process multiple transactions simultaneously, without violating the correctness of transactions. Increasing the possible number of concurrent transactions is an important technique to improve the throughput of transaction processing.

- **Autonomy:** Autonomous execution of mobile clients represents the ability to reduce the dependency upon the server while executing transactions. Autonomous execution of transactions at mobile clients can significantly enhance the performance, since it reduces the contention on the bandwidth of wireless link, which is scarce resource in mobile environments.

The design space of concurrency control protocols in mobile database systems is presented in this section, which presents a taxonomy of such protocols.

This taxonomy can be used to help understanding the design alternatives and performance tradeoffs for concurrency control algorithms.

The taxonomy is restricted to algorithms that are applicable to mobile client-server databases and that provide serializability as the correctness criterion. Algorithms that provide lower levels of correctness are not considered here. The taxonomy is shown in Fig. 3. There are a wide range of options for the design of such protocols. We present four levels of classification in the taxonomy, and at the top level of the taxonomy, algorithms are classified according to *the unit of certification* that they employ. This is a fundamental consideration, as it determines when a mobile client initiates the concurrency control action. Three more levels of lower classifications are presented in the taxonomy, but only one of them is applicable to the protocols whose unit of certification is an operation. For the protocols that are classified in the other group at the first level of taxonomy, all of three lower levels, *checking duration*, *local conflict action* and *read-only transaction* can be applied. Each level of the taxonomy is described in the next subsections.

3.1   Unit of Certification

Mobile clients send the certification request to the server, as transactions initiated at mobile clients should be checked by the server whether they satisfy the correctness criterion. Concurrency control protocols can thus be classified based on the unit of certification sent to the server. The protocols that have been proposed for mobile client-server database systems can be partitioned into two classes based on this criterion: *operation-based* and *transaction-based*.

From another point of view, the difference between the operation-based and transaction-based approaches lies in the type of operations a mobile client expects from the initiated transaction in future [1,2]. If a mobile client thinks that there will be conflicts related to the initiated transaction, it adopts an approach which prevents such conflicts by checking each operation of the transaction, thus it is also called pessimistic approach. On the other hand, if a mobile client expects that most of transactions will not conflict with others; that is there will be few sharing data items in the entire database, it just executes the entire transaction locally. The transaction is checked for consistency preservation only once at the server, thus it is called optimistic approach.

The operation-based scheme requires mobile clients to contact server for every operation of transactions [3,4,5,6]. The server checks if the operation can be processed without violating the correctness criterion, and notifies this information to the mobile client. Then the mobile client can process the next operation. When the mobile client completes all operations of a transaction, the mobile client can decide the commit of the transaction autonomously. This is because all the operations of the transaction have been verified by the server at the time of request.

In contrast, under the transaction-based scheme, mobile clients are not required to interact with the server while executing transactions [7,8,9,10,11]. Instead, when all operations of the transaction are completed, the mobile client sends the history of the executed transaction to the server in order to verify it. If the server replies that the transaction satisfies the correctness criterion as expected, the mobile client commits the transaction. Otherwise, the transaction should be aborted.

The main argument for the operation-based approach is that it can ensure correct execution. Because each operation of a transaction is checked separately when it is requested, it is impossible for a mobile client to execute any operation which does not satisfy the correctness criterion. Thus, once a transaction completes all the operations, it means that the execution of the transaction ensures the correctness criterion, and it can be committed immediately. As a result, no transaction produces a history which is not serializable. Thus, no transaction is aborted once it completes all operations. (Aborts can occur due to other reasons, such as deadlock between transactions or accessing stale cached data). The disadvantage of the operation-based scheme, however, is a greater dependency on the server. This can result in significant performance overhead, such as increased communication on wireless network.

Avoiding the drawback of the operation-based scheme is the main contribution of the transaction-based scheme. Because a mobile client executes a transaction autonomously until all operations are completed, the transaction-based scheme does not suffer from communication overhead which is fatal in mobile environments. However, this advantage of transaction-based scheme can be achieved at the cost of increased aborts of transactions. If the optimism which is the basis of the transaction-based scheme turns out to be unfounded, then the large portion of transactions which have been executed locally at mobile clients must be aborted. This can degrade the throughput of entire transaction processing.

3.2   Checking Duration

The second level of differentiation for the taxonomy is based on the duration for which the server maintains the checking information of each operation for a transaction (operation-based scheme), or on the point at which the server checks the correctness of transactions (transaction-based scheme).

First, in case of operation-based scheme, conflicting operations of different transactions cannot be granted to be executed concurrently by the server. As a result, the server should prevent two or more operations of conflicting transactions from accessing the same data concurrently. In the operation-based schemes, whenever an operation is requested by a mobile client, the server gives the right to access the data, which is mutually exclusive (e.g., locking). In the taxonomy, there are two classes of checking duration strategies for the operation-based schemes.

Immediate Release With this strategy, an access privilege of an operation is released immediately after the execution of the operation is completed. As a result, transactions can access a data which has been accessed by a conflicting operation of other transactions which are still active. The main advantage of the immediate release scheme is the higher throughput of the transactions due to the increased concurrency. Since transactions do not hold accessed data until they terminate, more transactions can be executed concurrently. However, in general, accessing data that are written by uncommitted transactions cannot guarantee the correct execution. [4] has proposed *speculative lock management* algorithm as the way of ensuring the correctness of the immediate release scheme. In this algorithm, a transaction releases locks on a data item whenever it writes corresponding data. The waiting transaction reads before- and after- images and carries out speculative execution. In order to satisfy the serializability, the transaction which has carried out speculative executions can commit only after termination of preceding transactions. On the termination of preceding transactions, it selects appropriate execution based on the termination decisions (i.e., commit or abort).

Delayed Release In the operation-based scheme, it is more general to release the access privilege when all the operations of a transaction are completed, and we classified such algorithms as delayed release scheme. With this scheme, once a privilege for a data item is acquired by a transaction, other transactions cannot access that data until the preceding transaction commits or aborts. Although the degree of concurrency is lower compared to the immediate release scheme, it is guaranteed that the schedule produced always ensures the correctness criterion. [5], [6] and [7] proposed concurrency control strategies that use locking with delayed release approach.

On the other hand, several transaction-based protocols have been proposed and studied in the literature. As mobile clients execute the entire transaction locally without any communication with the server, transactions should be checked for correctness before they terminate. Thus, a mobile client sends a messages to the server that request the commit of the transaction. When the server receives such a commit requesting message, it checks if the transaction satifies the correctness criterion. In the taxonomy, we can differentiate the transaction-based scheme into the following two classes according to the point at which the checking is performed.

Immediate Certification With this approach, whenever a mobile client requests commit of a transaction to a server, the server makes the decision of a transaction (i.e., commit or abort) *immediately after* the server receives commit requesting message [7,8,9]. The main argument for the immediate certification scheme is simplicity. Because certification actions of a transaction T_i in this case is performed against the transactions which have requested commit earlier than T_i, the server does not have to consider transactions whose commit requesting messages arrive after T_i. As a result, in this scheme, the order of arrival of commit requesting message is the main parameter which decides the commit or abort of the transaction.

Delayed Certification In the transaction-based scheme, the server can delay the decision of commit requesting transactions until the result of decision is actually sentsentsent to mobile clients. The delayed certification scheme was proposed and studied in [10], in order to increase concurrency of transactions. With the delayed certification process, the server can select transactions which are related to a large number of conflicts, and by aborting such transactions, the throughput can be improved. The delayed certification scheme can be applied when the server sends the results of certifications to mobile clients periodically.

3.3   Local Conflict Action

The next level of differentiation for the taxonomy is local conflict action, which is applicable only to the transaction-based scheme. This level is based on the priority given to the committing transactions at remote mobile clients to which they are sent. When the committing transaction shows a conflict with an active transaction at a mobile client, there are two options: preemption and non-preemption.

Preemption With the preemption scheme, ongoing transactions at mobile clients are aborted as the result of an incoming transaction which shows conflicts. Under this scheme, the optimism that is assumed in mobile clients regarding the execution of a transaction is somewhat weaker than under the non-preemption scheme. This is because the non-preemption scheme will force a committing transaction to serialize behind a locally ongoing transaction if a conflict is detected, whereas under the preemption scheme, committing transactions always have priority over ongoing transactions, so conflicting local transactions are aborted. When an active transaction which conflicts with committing transactions completes all operations, it can not be committed by the server, as it performed operations that conflict with already committed transactions. Thus, aborting such local transactions early in mobile clients can reduce unnecessary communication overhead. The preemption scheme was firstly proposed in [7], which proposed a concurrency control protocol called *Wound Certifier*. The key to the algorithm is the use of the broadcast channel to transmit information about read and write sets so mobile clients can decide whether their active transactions can continue or should be aborted. By checking in every period whether the current read and write set of the transaction intersects with those of committed transactions, the mobile client is acting as a *Wound Certifier* for its own transactions. The decision of aborting a transaction is done if the transaction's read or write set intersects with a committing transaction's read or write sets. [11] also has proposed two preemption schemes, versioning and invalidation method, which can abort active transaction at mobile clients using information broadcasted from the server. In this way, the algorithm downloads some of the work of validating transactions to the mobile clients, as a result, can increase the autonomy.

Non-preemption In contrast to the preemption scheme, with the non-preemption scheme, committing transactions does not preempt the ongoing conflicting transaction. Ongoing transactions continue their operations regardless of the committing transaction which conflicts. Then, most of such active transactions cannot be committed when they complete their operations. Although the procedure at mobile clients is simple with non-preemption scheme, it is quite wasteful to send a commit request message of a transaction that cannot be committed.

3.4   Commit of Read-only Transactions

When all operations of a transaction are read operations, we can consider a special commit process which is performed independently by mobile clients. We classify the concurrency control protocols according to the commit process of read-only transactions. This level of differentiation also is only applicable to the transaction-based scheme. In case of the operation-based scheme, as each operation should be guaranteed by the server, it is impossible to apply local commit policy.

Local Commit of Read-only Transactions As described in the previous section, using the transaction-based scheme, transactions executed locally at mobile clients must be sent to the server to be checked for the correctness. If mobile clients can decide commit or abort of a locally executed transaction only with information which is broadcasted from the server (without uplink message), the overall pathlength of transactions can be significantly shortened, and the throughput can be improved through the offloading of the wireless network. Although transactions which updated data items should be sent to the server because of the update installation, special consideration can be given to read-only transactions [10,11]. With the local commit policy for read-only transactions, a mobile client commits a transaction autonomously, if all operations of the transaction are read operations. Of course, there should be a special consideration to commit a read-only transaction locally, in order to ensure the correctness criterion.

Global Commit of Read-only Transactions If a protocol has no consideration for the local commit of read-only transactions, all transactions should be sent to the server to be guaranteed the correctness. Most of transaction-based protocols proposed in the literature adopted the global commit strategy for read-only transactions.

4. DATA DISSEMINATION

Traditionally, the mode of data delivery has largely been on request-response style. Users explicitly requests data items from the server. When a data request is received at a server, the server locates the information of interest and returns it to the client. This form of data delivery is called pull-based. In wireless computing environments, the stationary server machines are provided with a relatively high bandwidth channel which supports broadcast delivery to mobile devices in their cell. As a result, in recent years, different models of data delivery have been explored, particularly the periodic push-based model where servers repetitively broadcasts data to mobile clients without any explicit requests.

In mobile computing environments, several criteria can be used to evaluate the performance of a data delivery method.

- **Responsiveness:** The most important criterion of a data delivery scheme is its ability to get the requested data to the user quickly. In this regard, two metrics can be considered. The first one is the average access time, which is the amount of time spent, on average, from the instant the request is made to the time that the requested data item is received. The second metric is the worst-case access time that measures the maximum access time for any user request to be satisfied.

- **Scalability** In mobile client-server systems, one of the most important criteria is the cell capacity that measures the number of mobile devices which can be handled by a server. The effectiveness of a scheme is also determined by how well it adapts to workload or environmental changes. The scheme should be able to support increasingly large number of population of users.

- **Power Efficiency** As battery power is a precious resource for mobile devices, it has to be minimized.

- **Tuning Time** Another metric that is commonly used as an indication of the energy consumption of an data delivery strategy is the tuning time.

The tuning time measures the amount time that a mobile client listens to the channel. Thus, it measures the time during which the mobile client stays in the active mode and therefore determines the power consumed by the client to retrieve the relevant data.

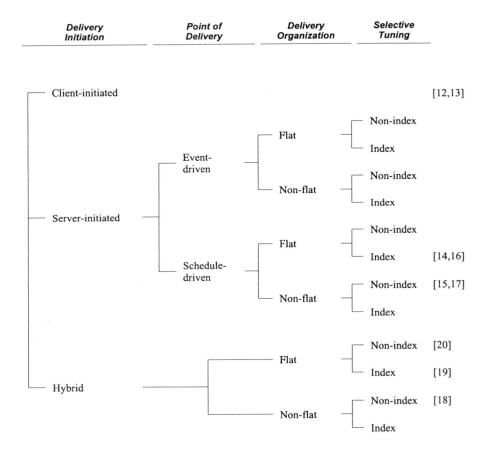

Figure 4. Taxonomy of data delivery schemes

We shall present an alternative taxonomy of data delivery mechanisms based on the design issues that need to be addressed in developing those schemes. This taxonomy is restricted to algorithms in the literature that were proposed for the mobile computing environments. The taxonomy is shown in Fig. 4. We present also four levels of the classification that can be broadly categorized as delivery initiation, point of delivery, delivery organization and selective tuning. The top level of classification is fundamental consideration, as it determines which part is obligated to initiate the delivery of data items. Three more levels of lower classifications are presented in the taxonomy, and all of them are applicable to the protocols whose delivery is initiated by the server. Each level of the taxonomy is described in the following.

4.1   Delivery Initiation

Data delivery can be initiated by the mobile client (client-initiated) or by the server (server-initiated). Under the client-initiated approach, mobile clients pull the desired data object from the server by submitting queries to the server, the server accesses the relevant data items from the database and returns them to the mobile clients. The server initiated approach operates in a different manner. The server initiates data delivery and sends data objects to mobile clients, possibly without any explicit requests from clients. In this case, to receive the data items, mobile clients must listen to the broadcast channel to filter the incoming stream for their desired data.

Client-initiated Delivery In traditional client-server systems, data are delivered on a demand basis. A client explicitly requests data items from the server. When a data request is received at a server, the server locates the data of interest and returns it to the mobile client. This form of data delivery is called client-initiated or pull based data delivery. The client-initiated approach is effective when the client population is small; otherwise, the communication channel and the server can become a bottleneck quickly. Moreover, it requires an uplink channel to be available and that the mobile client must have transmission capacity. In addition, mobile clients need to be able to know what they want to retrieve. The client-initiated data delivery is adopted in [12,13].

Server-initiated Delivery In the server-initiated approach, the responsibility of transferring data rests with the server. With this approach, the server delivers data items to mobile clients without any explicit request from it, and in anticipation of an access in the future. Thus, unlike in the pull based approach, the transfer is initiated by the server and not by the client, and this approach is called server-initiated approach or push based data delivery. The server has to have some knowledge of the client data requirements for this method to work well. Server-initiated approach is an effective mechanism for large client population, and avoids the limitations of the client-initiated schemes in mobile computing environments. However, it is also limited with a problem that it is difficult to predict accurately the needs of mobile clients. Sending irrelevant data results in poor use of the channel bandwidth, and data may not reach the mobile clients in time. The server-initiated data delivery has been explored in [14,15,16,17].

Hybrid Delivery Push and pull based delivery can be combined by considering systems in which besides the broadcast channel, mobile clients are provided with an uplink channel from the clients to the server, also called back channel, used to send messages to the server. This approach is called hybrid data delivery. In a hybrid approach, some data items are delivered by the server initiation, while the remaining data items are to be requested by mobile clients before they are delivered. One important factor in hybrid delivery is whether the same channel from the server to the clients is used for both broadcast delivery and for the transmission of the replies to on demand requests. In this case, techniques for efficiently sharing the channel are of interest. Mobile clients can use back channel to provide feedback and profile information to the server. They can also use the back channel to directly request data, for instance critical data for which they cannot wait to appear on the broadcast [18].

Instead of broadcasting all data items in the database, one way to realize hybrid data delivery is to partition data items into two sets: one being broadcasted and the other being available only on demand [19]. Determining which part of the database to broadcast is a complicated task since the decision depends on many factors including mobile clients' access patterns and the server's capacity to service requests.

Mobility of users is also critical in determining the set of broadcast items. Cells may differ in their type of communication infrastructure and thus in their capacity to service requests. Furthermore, as users move between cells, the distribution of requests for specific data at each cell changes. An adaptive algorithm that takes into account mobility of users between cells is proposed in [20].

4.2   Point of Delivery

This dimension of data delivery examines the schedulability of the data, i.e., whether the data items are delivered based on event-driven or schedule driven. This level of differentiation is only applicable to the server-initiated delivery scheme or the hybrid delivery scheme. In case of the client-initiates scheme, as the server delivers data items in response to the requests from mobile clients, it is impossible to apply schedule-driven policy.

Event-driven In data delivery scheme which adopts event-driven strategy, there is no predetermined schedule on how data items are to be delivered. Data items are disseminated in response to events such as requests or triggered by updates. Thus all the proposed client-initiated schemes are classified as event-driven data delivery.

Schedule-driven Data delivery schemes, which adopt schedule-driven policy, deliver data based on some predetermined schedule. For example, information may be sent out daily or weekly, or information may be polled periodically as in the remote-sensing application.

4.3 Delivery Organization

The data items to be delivered have to be organized for dissemination. The data items to be delivered may be organized in consideration of bandwidth utilization and performance improvements. Mobile clients are interested in accessing specific data items from the delivered data. The access time is the average time elapsed from the moment a mobile client expresses its interest to an item by submitting a query to the receipt of the item on the broadcast channel. The broadcasted data should be organized so that the access time is minimized. This level of classification is based on the strategy to organize the contents of information which is delivered by the server, and as a result, this lever is applicable to schedule-driven data delivery schemes. In the schedule-driven scheme, data items disseminated follows a regular, repeating program, and this program may be flat or non-flat.

Flat The simplest way to organize the transmission of broadcast data is a flat organization. In flat programs, all data items are of equal importance, and broadcasted once in a broadcast cycle [14,16]. Given an indication of the data items desired by each mobile client listening to the broadcast, the server simply takes the union of the required items and broadcast the resulting set cyclically. The regularity of a flat organization makes it easier to design mechanisms that allow mobile clients to search the desired portion of data.

Non-flat The basic idea of the non-flat organization is to broadcast data items that are most likely to be of interest to a larger part of the client community more frequently than others. Thus, the non-flat program favors objects with higher access frequencies. Hence, in a broadcast cycle of a non-flat organization, while all data items are broadcasted, some will appear more often than others. Doing so, in effect, creates an arbitrary fine grained memory hierarchy as the expected delay in obtaining an item depends on how often that item is broadcasted. Non-flat programs yields shorter access time for popular data items as compared to flat programs at the expense of longer access time for data items that are less frequently accessed. Non-flat programs also provide a better average access time than flat programs. However, it has higher bandwidth requirement as its broadcast cycle length is longer than than of flat programs. Non-flat strategies have been explored in [15,17].

4.4   Selective Tuning

In a server-initiated delivery mechanism, a mobile client listening to the channel needs to examine every data item that is broadcasted. The tuning time of data delivery is the amount of time spent listening to the broadcast channel. Listening to the broadcast channel requires the mobile client to be in the active mode and increase power consumption. Some mechanisms that adopted indexing have been proposed to minimize this power consuming process, which is scarce resource at mobile device, thus the last level of differentiation is selective tuning.

Non-index With the non-indexing data delivery mechanism, the server just broadcast data items in flat or non-flat organization. Mobile clients should then listen to the broadcast's channel until they obtain requested data items. This process requires the CPU to be in the active mode, which is a power consuming operation at mobile device. Since the mobile client is typically interested in only a small subset of the broadcasted data, the overhead of scanning the other objects is wasted.

Index Mobile clients may be interested in fetching from the broadcast individual data items identified by some key. To minimize the scarce energy resources in mobile devices, methods to index data have been proposed so that mobile clients only need to selectively tune to the desired data [14,16,19]. Thus, most of the time clients will remain in doze mode and thus save energy. The objective is to develop methods for allocating indeces together with data on the broadcast channel so that both access and tuning time are optimized.

5.   CACHE CONSISTENCY

The bandwidth of the wireless channel is rather limited, and as a result, *caching* of frequently accessed data in a mobile client can be an effective approach for reducing contention on the narrow bandwidth wireless channel. Caching allows the database systems to use the resource of mobile clients in order to reduce the number of data requests that must be sent to the server. The effectiveness of caching depends on the assumption that there is significant *locality of access* in the system workload. Locality can be considered along two dimensions:

- *Temporal Locality:* References to items are clustered in time. If an item is accessed, it is likely to be accessed again in the near future.

- *Spatial Locality:* References to items are clustered in space. If an item is accessed, it is likely that items that are physically near it will be referenced in the near future.

However, once caching is used, a *consistency maintenance* strategy is required to ensure the consistency of cached data. Because data items are allowed to be cached by multiple mobile clients, a mechanism for ensuring that all mobile clients see a consistent view of database should be used. This is referred to as the *cache consistency* problem. This is, unfortunately, difficult to enforce in a mobile computing environments due to the frequent disconnection and mobility of clients. Clients who resume connection no longer know whether their cached content is still valid.

Traditional techniques require either the server to transmit invalidation messages to the clients every time an object is updated or the mobile client to query the server to verify the validity of the cache contents. Both approaches, however, are not adequate for mobile computing environments. In the first approach, which has also been referred to as stateful based approach, the server must keep track of the mobile clients' cache content and locate the appropriate clients whenever a data item is updated. Moreover, even if a mobile client is not using a particular cached data item, it gets notified about its invalid status, which is a potential waste of bandwidth. In the second approach, the mobile client must send a message every time they want to use their cache. This is both wasteful of bandwidth and battery power of mobile devices.

This section provides a taxonomy of consistency maintenance protocols that encompasses the algorithms proposed in the literature. The taxonomy is restricted to algorithms that are applicable to mobile client-server database systems. In stateless-based cache consistency schemes, the server has no information about which clients are currently under its cell and what data items are cached by mobile clients. Most of cache consistency schemes proposed in the literature are stateless-based. The taxonomy is shown in Fig. 5. We present three levels of classification in the taxonomy, and at the top level, algorithms are classified according to by whom *the consistency action* is *initiated*. This is a fundamental consideration, as it determines which part is obligated to maintain the consistency. Each level of the taxonomy is described in the subsections that follow.

5.1 Consistency Action Initiation

Because cached data is replicated data, it follows that traditional methods for managing updates to replicated data can be used or extended to manage cached copies at mobile clients. Cache consistency maintenance protocols can thus be classified based on the result of a particular update. The protocols that have been proposed for mobile client-server databases can be partitioned into two classes based on this criterion: *client-initiated* and *server-initiated*.

From a qualitative point of view, the difference between the client-initiated and server-initiated approaches lies in how access to stale data is prevented.

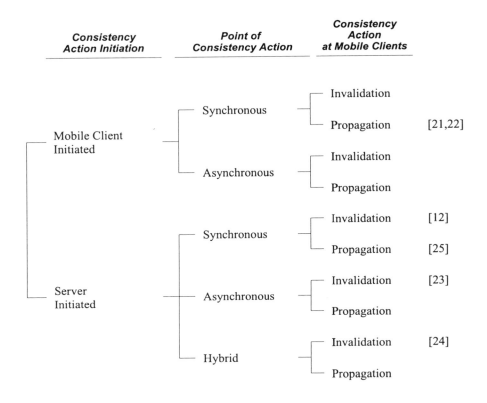

Figure 5.   Taxonomy of cache consistency schemes

Specifically, a data item is considered to be stale if its value is older than the item's latest committed value. Consider the concurrency control algorithms proposed in the previous section, in which data items are tagged with sequence numbers, and where the sequence number of an item is increased when a transaction that has updated the item commits. A copy of a data item is then considered to be stale, if its sequence number is lower than the latest sequence number assigned to any copy of the data item.

The *client-initiated* scheme allows stale data copies to reside in a mobile client's cache. The validity of cached data items are checked when they are actually used by a transaction [21,22]. As a result, mobile clients are responsible for checking the consistency of data. The client-initiated scheme is so named because the validity checking of cached data is explicitly performed by mobile clients.

In contrast, under the *server-initiated* scheme, the consistency of data copies at mobile clients is maintained by the server, with periodic or aperiodic broadcasting messages [12,23,24,25]. As a result, most of cached data at mobile clients are likely to have up-to-date value. However, the consistency of cached data is not strictly guaranteed, because there exists an interval for a committed updated transaction until it is notified to mobile clients.

The main argument for the client-initiated scheme is that mobile clients cannot execute transactions with stale data copies, as mobile clients check the validity of data whenever they are accessed by transactions. With the client-initiated scheme, stale data copies in mobile clients' cache are detected immediately during the execution of transactions, thus aborts can be avoided. The disadvantage of the client-initiated scheme, however, is increased communications between mobile clients and servers, especially uplink communications which occurs whenever cached copies are accessed. This can result in significant performance overhead, considering the asymmetric environment of wireless channel.

On the other hand, the server-initiated scheme does not require any uplink communication for the consistency check, although the consistency of cached data is not strictly guaranteed. Because the server is responsible for maintaining the consistency, mobile clients just listen for consistency information broadcasted from the server, instead of sending messages to check the validity. If a transaction accesses data which has been updated in the server. If the mobile client is not yet informed the update, the transaction will be aborted, as the inconsistency is detected at the end of the transaction.

5.2   Point of Consistency Action

The second level of differentiation for the taxonomy is based on the strategy which is adopted to check the validity of data items which are accessed by an active transaction (client-initiated scheme), or to send information about updated data items by the server (server-initiated scheme).

At first, the client-initiated scheme can be classified based on the time when mobile clients check the consistency of data touched by a transaction. The client-initiated scheme does not guarantee the consistency of data in mobile clients' cache, and as a result, the consistency of data items accessed by a transaction must be determined before the transaction can be allowed to commit. Thus, the consistency checks should begin and complete during the execution of a transaction. In the taxonomy, there are two classes of consistency checking strategies: *Synchronous* and *Asynchronous*.

Synchronous Checking On the first access that a transaction makes to a particular data item, the mobile client must communicate with the server to ensure that its copy of the item is valid [21,22]. In immediate checking scheme, this is done in a synchronous manner - the transaction is not allowed to access the data item until its validity has been verified. Once the validity of the mobile client's copy of the data has been established, the copy is guaranteed to remain valid at least until the transaction completes. The main argument of the immediate checking scheme is the strict consistency of accessed data. When a transaction completes all operations with the immediate checking scheme, it is always guaranteed that all data accessed by the transaction is valid, and as a result, can be committed. However, such guarantee of consistency is achieved at the cost of increased communications with the server.

Asynchronous Checking Delayed checking is an optimistic approach compared to the immediate checking. No consistency action is sent to the server until the transaction has completed its operations and is ready to commit. At this point, information on all the data items read and written by the transaction is sent to the server, and the server determines whether or not the transaction should be allowed to commit. This scheme can have two advantages. First, consistency actions can be bundled together in order to reduce consistency maintenance overhead. Secondly, any consistency maintenance work performed for a transaction that ultimately aborts is wasted; delayed checking can avoid some of this work. The main disadvantage of the scheme is, however, that the delay can result in the late detection of data conflicts. The resolution of inconsistency that are detected after they have occurred typically requires aborting one or more transactions.

On the other hand, in the server-initiated scheme, the server sends the consistency information to mobile clients by broadcasting messages, and in the taxonomy, we classified the server-initiated scheme into three classes based on the point at which such broadcasting messages are sent: *synchronous broadcasting, asynchronous broadcasting* and *hybrid broadcasting.*

Synchronous Broadcasting The synchronous broadcasting method is based on periodic broadcasting of invalidation reports. The server keeps track of the data items that are recently updated, and broadcast these information periodically. With this scheme, at a periodic broadcasting point, the server sends the list of data items which have been updated after the last broadcasting point. Since the broadcasting occurs periodically, the message overhead for the synchronous broadcasting scheme can be stable, and as a result, the communication overhead on wireless network is relatively low. However, it cannot adapt to the update frequency, and as a result, many cached copies at mobile clients can have inconsistent value, when frequent updates by transactions exist. A mobile client has to listen to the report to decide whether its cache is valid or not. Thus, each mobile client is confident for the validity of of its cache only as of the last invalidation report. That adds some latency to query processing, since to answer a query, a mobile client has to wait for the next invalidation report. This overhead in query latency can be avoided if a less strict consistency model is adopted. Three synchronous broadcasting strategies have been proposed in [12]. In the broadcasting timestamps strategy, the invalidation reports contain the timestamps of the latest change for data items updated in the last w seconds. In the amnestic terminals strategy, the server only broadcasts the identifiers of data items changed since the last invalidation report. In the signature strategy, signatures are broadcasted. A signature is a checksum computed over the value of a number of data items by applying compression technique.

Asynchronous Broadcasting The consistency information can be broadcasted immediately after changes to data items occur, in order to adapt to update frequency by transactions. This approach is called the asynchronous broadcasting scheme, which is proposed in the following section. With the asynchronous broadcasting scheme, the server can adjust the broadcasting period according the frequency of updates on data items, and as a result, is effective for connected mobile clients, and allows them to be notified immediately of updates. However, for a mobile client who reconnects after a period of disconnection, the client has no idea of what has been updated and so the entirety of its cache content has to be invalidated. An asynchronous technique based on bit sequences has been proposed in [23]. In this strategy, the invalidation report is organized as a set of bit sequences with an associated set of timestamps. Each bit in the sequence represents a data item in the database. The set of bit sequences is organized in a hierarchical structure. It is shown that the algorithm performs consistently well under conditions of variable update rate and client disconnection time.

Hybrid Broadcasting The hybrid broadcasting scheme is an hybrid approach between synchronous and asynchronous broadcasting. In order to adapt to the caching pattern of each data item, this scheme selects asynchronous broadcasting for widely cached data items. and synchronous broadcasting for exclusively used data items [24]. As a result, this scheme can reduce communication overhead which is indispensable for the asynchronous broadcasting, while still adapting to the update frequency.

5.3 Consistency Action at Mobile Clients

The next level of differentiation for the taxonomy is consistency action which is performed at mobile clients when they receive the broadcasted consistency information. This level of differentiation is applicable only to the server-initiated schemes, and there are two options here: *invalidation* and *propagation*.

Invalidation With the invalidation scheme, when mobile clients receive the list of updated data items which is broadcasted from the server, they remove the stale copy of data items from the cache, so it will not be accessed by any subsequent transactions. After a data item is invalidated at a mobile client, a subsequent transaction that wishes to access the data item at that mobile client must obtain a new copy from the server. The invalidation scheme is adopted in most of cache consistency algorithms proposed for mobile environments [12,22,23,24].

Propagation Propagation results in the newly updated value being installed at the mobile client in the place of stale copy [21,22,25]. In this way, mobile clients do not have to request newly updated data to the server. However, transmitting the updated value of data items may be an overhead in terms of wireless communication. Thus, most of propagation schemes are proposed for the traditional wired client-server database systems. There is a tradeoff between invalidation and propagation. Under the propagation strategy, when disconnection time is short, mobile clients can update their cache immediately. Under invalidation strategy, mobile clients must still submit requests to retrieve the updated records even if the disconnection is short. However, under the propagation strategy, since the entire content of a data item is broadcasted, the report is much larger and can take up a significant portion of downlink channel capacity which is a scarce resource in mobile computing environments.

6.   CONCLUSIONS

This chapter investigated a range of data management techniques which utilize broadcasting facility in mobile computing environments. Three main research topics were addressed: concurrency control, data dissemination and cache consistency. Broadcasting approach to transmit information to numerous concurrent mobile clients is attractive in mobile computing environment, because a server need not know the location and the connection status of its clients, and because the clients need not establish an uplink connection which is expensive in asymmetric communication environment. In this chapter, we presented design spaces of various algorithms proposed in those research topics. These taxonomies can be used to help understanding the design alternatives and performance tradeoffs of those algorithms.

REFERENCES

[1]   R.E. Gruber, *"Optimism vs. Locking: A Study of Concurrency Control for Client-Server Object-Oriented Databases,"* Ph.d. Thesis, Dept. of

Electrical Engineering and Computer Science, Massachusetts Institute of Technology, 1997.

[2] I. Chung, J. Lee and C.-S. Hwang, "A Contention Based Dynamic Consistency Maintenance for Client Cache," in *Proc, of International Conference on Information and Knowledge Management*, 1997, pp. 363-370.

[3] J. Jing, O. Bukhres and A. Elmgamid, "Distributed Lock Management for Mobile Transactions," in *Proc. of IEEE International Conference on Distributed Computing Systems*, 1995, pp. 118-125.

[4] P. Reddy and M. Kitsuregawa, "Speculative Lock Management to Increase Concurrency in Mobile Environments," in *Proc. of International Conference on Mobile Data Access, Lecture Note in Computer Science*, vol. 1748, Springer, 1999, pp. 82-96.

[5] Q. Lu and M. Satyanarayanan, "Resouese Conservation in a Mobile Transaction System," *IEEE Transactions on Computer*, vol. 46, no. 3, 1997, pp. 299-311.

[6] A.K. Elmagarmid, J. Jing and O.A. Bukhres, "An Efficient and Reliable Reservation Algorithm for Mobile Transactions," in *Proceedings of International Conference on Information and Knowledge Management*, 1995, pp. 90-95.

[7] D. Barbara, "Certification Reports: Supporting Transactions in Wireless Systems," in *Proceedings of IEEE International Conference on Distributed Computing Systems*, 1997, pp. 466-473.

[8] V.C.S. Lee and K.-W. Lam, "Optimistic Concurrency Control in Broadcast Environments: Looking Forward at the Server and Backward at the Clients," in *Proceedings of International Conference on Mobile Data Access, Lecture Note in Computer Science*, vol. 1748, Springer, 1999, pp. 97-106.

[9] J. Shanmugasundaram, A. Nithrakashyap and R. Sivasankaran, "Efficient Concurrency Control for Broadcast Environments," in *Proceedings of ACM SIGMOD International Conference on Management of Data*, 1999, pp. 85-96.

[10] I. Chung and C.-S. Hwang, "Increasing Concurrency of Transactions using Delayed Certification," in *Proceedings of International Conference on Mobile Data Management*, 2001, pp. 277-278.

[11] E. Pitoura and P.K. Chrysanthis, "Exploiting Versions for Handling Updates in Broadcast Disks," in *Proceedings of International Conference on Very Large Databases*, 1999, pp. 114-125.

[12] D. Barbara and T. Imielinsky, "Sleepers and Workaholics: Caching Strategy in Mobile Environments," *VLDB Journal*, vol.4, no.4, 1995, pp. 567-602.

[13] K.L. Tan and B.C. Ooi, "Batch Scheduling for Demand-driven Servers in Wireless Environments," *Information Sciences*, vol. 109, 1998, pp.281-198.

[14] T. Imielinski, S. Viswanathan and B.R. Badrinath, "Energy Efficient Indexing on Air," in *Proceedings of ACM SIGMOD International Conference on Management of Data*, 1994, pp. 25-36.

[15] N. Vaidya and S. Hameed, "Scheduling Data Broadcast in Asymmetric communication environments," *ACM/Baltzer Wireless Networks*, vol. 5, no. 3, 1999, pp. 171-182.

[16] E. Pitoura and P.K. Chrysanthis, "Scalable Processing of Read-Only Transactions in Broadcast Push," in *Proceedings of International Conference on Distributed Computing Systems*, 1999, pp. 432-439.

[17] S. Acharya, R. Alonso, M.J. Franklin and S.B. Zdonik, "Broadcast Disks: Data Management for Asymmetric Communications Environments," in *Proceedings of ACM SIGMOD International Conference on Management of Data*, 1995, pp. 199-210.

[18] S. Acharya, M. Franklin and S. Zdonik, "Balancing Push and Pull for Data Broadcast," in *Proceedings of ACM SIGMOD International Conference on Management of Data*, 1997, pp. 183-194.

[19] K. Stathatos, N. Roussopoulos and J.S. Baras, "Adaptive Data Broadcast in Hybrid Networks," in *Proceedings of International Conference on Very Large Data Bases*, 1997, pp. 326-335.

[20] A. Datta, A. Celik, J. Kim, D. Vander and V. Kumar, "Adaptive Broadcast Protocols to Support Efficient and Energy Conserving Retrieval from Databases in Mobile Computing Environments," in *Proceedings of International Conference on Data Engineering*, pp.124-133, 1997.

[21] M.H. Wong and W.M. Leung, "A Caching Policy to Support Read-only Transactions in a Mobile Computing Environment," Technical Report, Dept. of Computer Science, The Chinese Univ. of Hong Kong, 1995.

[22] W.-C. Peng and M.-S. Chen, "A Dynamic and Adaptive Cache Retrieval Scheme for Mobile Computing," in *Proceedings of IFCIS International Conference on Cooperative Information Systems*, 1998, pp. 251-259.

[23] J. Jing, A. Elmagarmid, A. Helal and A. Alonso, "Bit Sequences: An Adaptive Cache Invalidation Method in Mobile Client/Server Environments," *Mobile Networks and Applications*, vol. 2, no. 2, 1997, pp. 115-127.

[24] I. Chung, J. Ryu and C.-S. Hwang, "Efficient Cache Management Protocol Based on Data Locality in Mobile DBMSs," in *Current Issues in Databases and Information Systems, Proceedings of Conference on Advances in Databases and Information Systems, Lecture Note in Computer Science*, vol. 1884, Springer, 2000, pp. 51-64.

[25] J. Cai, K.L. Tan and B.C. Ooi, "On Incremental Cache Coherency Schemes in Mobile Computing Environment," in *Proceedings of International Conference on Data Engineering*, 1997, pp. 114-123.

Chapter 8

DATA AND TRANSACTION MANAGEMENT IN A MOBILE ENVIRONMENT

Sanjay Madria, Bharat Bhargava, Mukesh Mohania, Sourav Bhowmick
Department of Computer Science, University of Missouri-Rolla, USA; Department of Computer Sciences, Purdue University, USA; Department of Computer Science, Western Michigan University, USA; School of Computer Engineering, Nanyang Technological University, Singapore

Abstract: The mobile computing paradigm has emerged due to advances in wireless or cellular networking technology. This rapidly expanding technology poses many challenging research problems in the area of mobile database systems. Mobile users can access information independent of their physical location through wireless connections. However, accessing and manipulating information without restricting users to specific locations complicates data processing activities. There are computing constraints that make mobile database processing different from the wired distributed database computing. In this chapter, we survey the fundamental research challenges particular to mobile database computing, review some of the proposed solutions and identify some of the upcoming research challenges. We discuss interesting research areas, which include mobile location data management, transaction processing and broadcast, cache management and replication. We highlight new upcoming research directions in mobile digital library, mobile data warehousing, mobile workflow and mobile web and e-commerce.

Keywords: Distributed database, Mobile computing, Mobile database, Wireless or cellular networking

1. INTRODUCTION

The rapid technological advancements in cellular communications, wireless LAN and satellite services have led to the emergence of mobile computing [1]. In mobile computing, users are not attached to a fixed geographical location, instead their point of attachment to the network changes as they move. The emergence of relatively sophisticated low-power, low-cost and portable computing platforms such as laptops and personal digital assistants (PDA) have made it possible for people to work from anywhere at any time (from their offices, homes, and while travelling) via a wireless communication network to provide unrestricted user mobility.

Mobility and portability pose new challenges to mobile database management and distributed computing [2]. There is a necessity to design specifications for energy efficient data access methodologies and in general develop database software systems that extend existing database systems designs and platforms to satisfy the constraints imposed by mobile computing (see Figure 1). How to handle long periods of disconnection, and other constrained resources of mobile computing such as limited battery life and variable bandwidth etc.? In mobile computing, there will be more competition for shared data since it provides users with the ability to access information and services through wireless connections that can be retained even while the user is moving. Further, mobile users will have to share their data with others. The task of ensuring consistency of shared data becomes more difficult in mobile computing because of limitations and restrictions of wireless communication channels.

• Low bandwidth.
• Frequent disconnections, but predictable.
• High bandwidth variability.
• Expensive.
• Broadcast is physically supported in a cell.
• Limited battery power.
• Small size and screen of laptop.
• Susceptible to damaging data due to theft and accidents.
• Fast changing locations.
• Scalability
• Security

Figure 1. Constraints of mobile computing

In this chapter, we discuss some of the problems identified with mobile database computing, review proposed solutions, and explore the upcoming research challenges.

The rest of the chapter is as follows: In Section 2, we discuss mobile database architecture. Section 3 contains mobile data management research issues. In Section 4, we discuss transaction management issues in mobile computing. Section 5 presents some research directions in mobile data management, and the last section concludes the chapter.

2. MOBILE DATABASE ARCHITECTURE

In a mobile computing environment (see Figure 2), the network consists of Fixed Hosts (FH), Mobile Units (MU) and Base Stations (BS) or Mobile Support Stations (MSS). MUs are connected to the wired network components only through BS via wireless channels. MUs are battery powered portable computers, which move around freely in a restricted area, which we refer to as the "geographical region" (G). For example in Figure 2, G is the total area covered by all BS. This cell size restriction is mainly due to the limited bandwidth of wireless communication channels. To support the mobility of MUs and to exploit frequency reuse, the entire G is divided into smaller areas called cells. Each cell is managed by a particular BS. Each BS will store information such as user profile, log-in files, access rights together with user's private files. At any given instant, a MU communicates only with the BS responsible for its cell. The mobile discipline requires that a MU must have unrestricted movement within G (inter-cell movement) and must be able to access desired data from any cell.

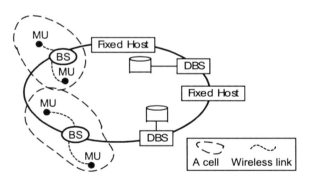

Figure 2. Architecture of MDS

A mobile unit (MU) changes its location and network connections while computations are being processed. While in motion, a mobile host retains its network connections through the support of base stations (BS) with wireless connections. The BSs and FHs (fixed hosts) perform the transaction and data management functions with the help of a database server (DBS) component to incorporate database processing capability without affecting any aspect of the generic mobile network. DBSs can either be installed at BSs or can be a part of FHs or can be independent of BS or FH.

When a MU leaves a cell serviced by a BS, a hand-off protocol is used to transfer the responsibility for mobile transaction and data support to the BS of the new cell. This hand-off involves establishing a new communication link. It may also involve migration of in progress transactions and database states from one BS to another. The entire process of handoff is transparent to a MU and is responsible for maintaining end-to-end data movement connectivity.

2.1   Modes of Operations

In mobile computing, there are several possible modes of operations [3,4], whereas in a traditional distributed system, a host may operate only in one of two modes; either connected to the network or totally disconnected. The operation mode in mobile computing may be one of the following:

- fully connected (normal connection)
- totally disconnected (e.g., not a failure of MU)
- partially connected or weak connection (a terminal is connected to the rest of the network via low bandwidth).

In addition, for conserving energy, a mobile computer may also enter an energy conservation mode, called *doze state* [3]. A doze state of MU does not imply the failure of the disconnected machine. In this mode, the clock speed is reduced and no user computation is performed.

Most of these disconnected modes are predictable in mobile computing. Protocols can be designed to prepare the system for transitions between various modes. A mobile host should be able to operate autonomously even during total disconnection.

A *disconnection protocol* is executed before the mobile host is physically detached from the network. The protocol should ensure that enough information is locally available (cached) to the mobile host for its autonomous operation during disconnection. It should inform the interested parties about the forthcoming disconnection.

A *partially-disconnection protocol* prepares the mobile host for operation in a mode where all communication with the fixed network is restricted. Selective caching of data at the host site will minimize future network use.

Recovery protocols re-establish the connection with the fixed network and resume normal operation.

Similar terminology has been used later.

3.   MOBILE DATA MANAGEMENT

In this section, we will discuss some of the important data management issues with respect to mobile computing.

3.1   Location Data Management

The location of a mobile user is of prime importance in wireless computing. In mobile computing, the location of a user can be regarded as a data item whose value changes with every move. In mobile computing, the location management is a data management problem; that is, the managing location data encounters the same problems as in managing normal data. Primary issues here are how to know the current position of the mobile unit. Where to store the location information and who should be responsible for determining and updating of information? To locate users, distributed location databases are deployed which maintain the current location of mobile users. The location data can be treated as a piece of data that is updated and queried. The search of this piece of data should be as efficient as any other queried data. Writing the location variable may involve updating the location of the user in the location database as well as in other replicated databases. The location management involves, searching, reading, informing and updating. If A wants to find the location of B, should A search the whole network or only look at pre-defined locations? Should B inform any one before relocating? One such method is described in [5]. It assumes that each user is attached to a home location server (now generally referred as home location register (HLR)) that always "knows" his current address. When a user moves, he informs his home location server about his new address. To send a message to such a user, his home location register is contacted first to obtain his current address. A special form of "address embedding" is used to redirect the packets addressed to the mobile user from the home location to his current location. This scheme works well for the user who stays within their respective home areas, it does not work for global moves. In this algorithm, when a user A calls user B, the lookup algorithm initiates a remote lookup query to the HLR of B, which may be at remote site. Performing remote queries can be slow due to high network latency. An improvement over such algorithm [6] is to maintain visitor location registers (VLR). The VLR at a geographical area stores the profiles of users currently located in that area for whom the area is not their home. The query then calls in the caller's area and if the callee's profile is not found, it queries the database in the callee's home area. This is useful when a callee received many calls from users in the area he is visiting since it avoids queries to the HLR of the callee at the remote site. VLR's can be viewed as a limited replication scheme since each user's profile is located in its current area when he is not in his home area. Another scheme proposed in [7], handles global moves on the assumption that most messages are exchanged between parties or between users in a remote area and its home location area.

A formal model for online tracking of users is considered in [8] by decomposing a PCS (Personal Communication System) network into regions and using regional directories. The authors discuss how to trade-off search and update costs while tracking users. Authors in [9,10] propose per-user placement, which uses cell partitions where the user travels frequently and separating the cells between which it relocates infrequently to control network traffic generated by frequent updates. Only moves which are across the partitions, are reported.

In [11], the location lookup problem is considered to find a callee within the reasonable time bounds to set up the call from the caller to callee. Each user is located in some geographical area where the mobile service station keeps track of each user in the form of <PID, ZID> where PID and ZID uniquely identified the mobile unit id and its current location id, respectively. They replicate per-user profiles based on calling and mobility patterns. Thus, they balance the storage and update costs and at the same time provide fast lookups. The decision of where to replicate the profiles is based on a minimum-cost maximum-flow [12] algorithm. They maintain an up-to-date copy of a user profile at the user's HLR and in addition, they also find out the sites at which a user's profile will be replicated. Thus, the algorithm does not guarantee that a user's profile will be found in his current area.

Hierarchical distributed database architectures [13,14,15,16,17] have been built to accommodate the increased traffic associated with locating moving users. In these models, each leaf database covers a specific geographical region and maintains location information for all users currently residing in that region. Location databases at internal nodes contain information about the location of all the users registered in areas covered by the databases at their children nodes. A hierarchical method for location binding in a wide-area system is used in the Globe wide-area location service [14]. Globe uses a combination of caching and partitioning.

A tree based structure is used for a location database in [18]. The authors modify the structure to balance the average load of search requests by replacing the root and some of the higher levels of the tree with a set-ary butterfly (a generalization of K-ary butterfly). They modify the lowest level of the tree to reflect neighbouring geographical regions more accurately and to allow simple hand-offs.

In [19], the hierarchical scheme allows dynamic adjustment of the user location information distribution, based on mobility patterns of mobile units.

A unique distribution strategy is determined for each mobile terminal and location pointers are set up at selected remote locations. This reduces database access overhead for registration and there is no need for centralized co-ordination.

Forwarding pointers have been used in hierarchical location databases [15,17]. In [15], the objective is to reduce the cost of moves by updating only databases up to a specific level of the tree and a forwarding pointer is set at a lower level in the database. However, if forwarding pointers are never deleted, then long chains are created, whose traversal results increases the cost of locating users during calls. They introduced caching techniques that reduce the number of forwarding pointers to travel before locating the calling as well as conditions for initiating a full update of the database entries. They have also described a synchronization method to control the concurrent execution of call and move operations. The difference between the two schemes [15,17] is that in [17] the actual location is saved at each internal level database instead of pointer to the corresponding lower level database. The forwarding pointers are set at different levels in the hierarchy and not necessarily at the lower level database as in [15]. In [17], the objective is to choose an appropriate level for setting the forwarding pointers and on updating obsolete entries in the hierarchy after a successful call. Caching in hierarchical structures is proposed in [13] instead of replication to reduce the cost of calls.

3.2 Cache Consistency

Caching of frequently accessed data plays an important role in mobile computing because of its ability to alleviate the performance and availability limitations during weak-connections and disconnections. Caching is useful during frequent relocation and connection to different database servers. In wireless computing, caching of frequently accessed data items is an important technique that will reduce contention on the small bandwidth wireless network. This will improve query response time, and to support disconnected or weakly connected operations. If a mobile user has cached a portion of the shared data, he may request different levels of cache consistency. In a strongly connected mode, the user may want the current values of the database items belonging to his cache. During weak connections, the user may require weak consistency when the cached copy is a quasicopy of the database items. Each type of connection may have a different degree of cache consistency associated with it. That is, a weak connection corresponds to a "weaker" level of consistency.

Cache consistency is severely hampered by both the disconnection and mobility of clients since a server may be unaware of the current locations and connection status of clients. This problem can be solved by the server by periodically broadcasting either the actual data, invalidation report (reports the data items which have been changed), or even control information such as lock tables or logs. This approach is attractive in mobile environments since the server need not know the location and connection status of its clients and clients need not establish an up link connection to a server to invalidate their caches. There are two advantages of broadcasting. First, the mobile host saves energy since they need not transmit data requests and second, broadcast data can be received by many mobile hosts at once with no extra cost.

Depending upon what is broadcasted, the appropriate schemes can be developed for maintaining consistency of data of a distributed system with a mobile client. Given the rate of updates, the trade-off is between the periodicity of broadcast and divergence of the cached copies that can be tolerated. The more the inconsistency is tolerated the less often the updates need to be broadcasted. Given a query, the mobile host may optimise energy costs by determining whether it can process the query using cached data or transmit a request for data. Another choice could be to wait for the relevant broadcast.

Cache coherence preservation under weak-connections is expensive. Large communication delays increase the cost of validation of cached objects. Unexpected failures increase the frequency of validation since it must be performed each time communication is restored. An approach that only validates on demand could reduce validation frequency but this approach would worsen consistency since it increases the possibility of some old objects being accesses while disconnected.

In Coda [20], during the disconnected operation, a client continues to have read and write access to data in its cache. The Coda file system allows the cached objects in the mobile host to be updated without any co-ordination. When connectivity is restored, the system propagates the modifications and detects update conflicts. The central idea is that caching of data and the key mechanisms for supporting disconnected operations which includes three states: hoarding, emulation and reintegration. The client cache manager while in hoarding state relies on server replication, but is always on the alert for possible disconnection and ensures that critical objects are cached at the time of disconnection. Upon disconnection, it enters the emulation state and relies solely on the contents of the cache. Coda's original technique for cache coherence while connected was based on callbacks [21]. In this technique, a server remembers that a client has cached an object, and promises to notify it when another client updates the object. This promise is called callback, and the invalidation message is a callback break. When a callback break is received, the client discards the cached copy and refetches it on demand. When a client is disconnected, it can no longer rely on callbacks. Upon reconnection, it must revalidate all cached objects before use to detect updates at the server.

Cache invalidation strategies will be affected by the disconnection and mobility of clients. The server may not have information about the live mobile units in its cells. [22] proposes taxonomy of different cache invalidation schemes and studies the impact of a client's disconnection times on their performance. They address the issue of relaxing consistency of caches. They use quasi-copies whose values can deviate in a controlled manner. They have categorized the mobile units on the basis of the amount of time they spend in their sleep mode into sleepers, and workaholics. Different caching schemes turn out to be effective for different populations. Broadcast with timestamps are proved to be advantageous for frequent queries than the rate of updates provided that units are not workaholics.

A technique is proposed in [23] to decide whether the mobile unit can still use some items in the cache even after it is connected to the server. The database is partitioned into different groups and items in the same group are cached together to decrease the traffic. Thus, a mobile unit has to invalidate only the group rather than individual items. In [24], various alternative caching strategies for mobile computing have been evaluated. More work needs to be done in the direction of performance evaluation and the availability limitation of various caching under weakly-connected and disconnected operation.

In [25], an incremental cache coherency problem in mobile computing is examined in the context of relational operations select, project and join. A taxonomy of cache coherency schemes is proposed as case studies. However, it does not address the problems of query processing and optimisation and does not include other relational operations.

3.3   Data Replication

The ability to replicate the data objects is essential in mobile computing to increase availability and performance. Shared data items have different synchronization constraints depending on their semantics and particular use. These constraints should be enforced on an individual basis. Replicated systems need to provide support for disconnected mode, data divergence, application defined reconciliation procedures, optimistic concurrency control, etc. Replication is a way by which the system ensures transparency for mobile users. A user who has relocated and has been using certain files and services at the previous location wants to have his environment recreated at the new location. Mobility of users and services and its impact on data replication and migration will be one of the main technical problems to be resolved. There are many issues raised by the relocated data and mobility of users and services:

- How to manage data replication, providing the levels of consistency, durability and availability needed?
- How to locate objects of interest? Should information about location be also replicated and to what extent (location is dynamically changing data item)?
- What are the conditions under which we need to replicate the data on a mobile site?
- How users' moves affect the replication scheme. How should the copy follow the user? In general data should move closer to the user?
- Is a mobile environment requiring dynamic replication schemes [26]?
- Do we need new replication algorithms or the proposed replication schemes for distributed environment can be modified?

In [27], caching of data in mobile hosts and the cost of maintaining consistency among replicated data copies have been discussed. It allows caching of data to take place anywhere along the path between mobile/fixed servers and clients. It determines, via simulations, which caching policy best suits given mobility and read/write patterns.

A general model is considered in [28] for maintaining consistency of replicated data in distributed applications. A casualty constraint, a partial ordering between application operations, is defined such that data sharing is achieved by defining groups requiring it and broadcasting updates to the group. Each node processes the data according to the constraints.

In [29], it has been argued that traditional replica control methods are not suitable for mobile databases and the authors have presented a virtual primary copy method. In this method, the replica control method decides on a transaction-by-transaction basis whether to execute that transaction on a mobile host's primary copy or virtual primary copy. This method requires a transaction to be restarted when a mobile host disconnects. Also, when a mobile host reconnects, it either has to wait for the completion of all transactions executed on a virtual copy before synchronizing itself with the rest of the system or all running transactions will have to be restarted.

An analysis of various static and dynamic data allocation methods is presented in [30] with the objective of optimising the communication cost between a mobile computer and the stationary computer that stores an online database. The authors consider one-copy and two-copies allocation schemes. In the static scheme, an allocation remains unchanged where as in a dynamic scheme allocation method changes are based on the number of reads and writes. If in the last k requests there are more reads at an MU than writes at a stationary computer, it uses the two-copy scheme. Otherwise it uses one-copy schemes. Two costs models were developed for cellular phones (user is charged per minute of connection) and packet radio networks (user is charged per message basis), respectively.

A new two-tier replication algorithm is proposed by Gray et al. in [31] to alleviate the unstable behaviour observed in the update anywhere-anytime-anyway transactional replication scheme when the workload scales up. Lazy master replication that is employed in the algorithm assigns an owner to each object. The owner stores the object's correct value. Updates are first done by the owner and then propagated to other replicas. The two tier scheme uses two kinds of nodes: mobile nodes (may be disconnected) and base nodes (always connected). The mobile nodes accumulate tentative transactions that run against the tentative database stored at the node. Each object is mastered at either the mobile node or the base node. When the mobile node reconnects to the base station, it sends replica updates mastered at the mobile node, the tentative transactions and their input parameters to the base node. They are to be re-executed as base transactions on the master version of data objects maintained at the base node in the order in which they are committed on the mobile node. If the base transaction fails its acceptance criterion, the base transaction is aborted and a message is returned to the user of the mobile node. While the transaction executed on the objects mastered on the mobile nodes are confirmed, those executed on the tentative objects have to be checked with nodes that hold the master version.

A dynamic replication scheme which employs user profiles for recording users' mobility pattern, access behaviour and read/write patterns, also actively reconfigures the replicas to adjust to the changes in the user's locations and systems is proposed in [32]. They devise the concept of open objects to represent a user's current and near future data requirements. This leads to a more precise and responsive cost model to reflect changes in access patterns.

4. MOBILE TRANSACTION PROCESSING

A transaction in mobile environment is different than the transactions in the centralised or distributed databases in the following ways:
- The mobile transactions might have to split their computations into sets of operations, some of which execute on a mobile host while others execute on stationary host.
- A mobile transaction shares its states and partial results with other transactions due to disconnection and mobility.
- The mobile transactions require computations and communications to be supported by stationary hosts.

- When the mobile user moves during the execution of a transaction, it continues its execution in the new cell. The partially executed transaction may be continued at the fixed local host according to the instruction given by the mobile user. Different mechanisms are required if the user wants to continue the transaction at a new destination.
- As the mobile hosts move from one cell to another, the states of transaction, states of accessed data objects, and the location information also move.
- The mobile transactions are long-lived transactions due to the mobility of both the data and users, and due to the frequent disconnections.
- The mobile transactions should support and handle concurrency, recovery, disconnection and mutual consistency of the replicated data objects.

To support mobile transactions, the transaction processing models should accommodate the limitations of mobile computing, such as unreliable communication, limited battery life, low band-width communication, and reduced storage capacity. Mobile computations should minimise aborts due to disconnection. Operations on shared data must ensure correctness of transactions executed on both the stationary and mobile hosts. The blocking of a transaction's executions on either the stationary or mobile hosts must be minimized to reduce communication cost and to increase concurrency. Proper support for mobile transactions must provide for local autonomy to allow transactions to be processed and committed on the mobile host despite temporary disconnection.

Semantic based transaction processing models [33,34] have been extended for mobile computing in [35] to increase concurrency by exploiting commutative operations. These techniques require caching a large portion of the database or maintain multiple copies of many data items. In [35], fragmentability of data objects have been used to facilitate semantic based transaction processing in mobile databases. The scheme fragments data objects. Each fragmented data object has to be cached independently and manipulated synchronously. That is, on request, a fragment of a data object is dispatched to the MU. On completion of the transaction, the mobile hosts return the fragments to the BS. Fragments are then integrated in the object in any order and such objects are termed as reorderable objects. This scheme works only in the situations where the data objects can be fragmented like sets, stacks and queues.

In optimistic concurrency control based schemes [36], cached objects on mobile hosts can be updated without any co-ordination but the updates need to be propagated and validated at the database servers for the commitment of transactions. This scheme leads to aborts of mobile transactions unless the conflicts are rare. Since mobile transactions are expected to be long-lived due to disconnection and long network delays, the conflicts will be more in a mobile computing environment.

In pessimistic schemes in which cached objects can be locked exclusively, mobile transactions can be committed locally. The pessimistic schemes lead to unnecessary transaction blocking since mobile hosts can not release any cached objects while it is disconnected. Existing caching methods attempt to cache the entire data objects or in some case the complete file. Caching of these potentially large objects over low bandwidth communication channels can result in wireless network congestion and high communication cost. The limited memory size of the MU allows for only a small number of objects to be cached at any given time.

Dynamic object clustering has been proposed in mobile computing in [4]. It assumes a fully distributed system, and the transaction model is designed to maintain the consistency of the database. The model uses weak-read, weak-write, strict-read and strict-write. The decomposition of operations is done based on the consistency requirement. Strict-read and strict-write have the same semantics as normal read and write operations invoked by transactions satisfying ACID (Atomicity, Consistency, Isolation and Durability) properties. A weak-read returns the value of a locally cached object written by a strict-write or a weak-write. A weak-write operation only updates a locally cached object, which might become permanent on cluster merging if the weak-write does not conflict with any strict-read or strict-write operation. The weak transactions use local and global commits. The local commit is the same as pre-commit of [37] and global commit is the same as a final commit in [37]. However, a weak transaction after local commit can abort and is compensated. In [37], a pre-committed transaction does not abort, hence requires no undo or compensation. A weak transaction's updates are visible to other weak transactions whereas prewrites are visible to all transactions.

A new transaction model using isolation-only transactions (IOT) is presented in [38]. The model supports a variety of mechanisms for automatic conflict detection and resolution. IOTs are sequences of file accesses that unlike traditional transactions have only isolation property. Transaction execution is performed entirely on the client and no partial result is visible on the servers. IOTs do not provide failure atomicity, and only conditionally guarantee permanence. They are similar to the weak transactions of [3].

An open nested transaction model has been proposed in [39] for modelling mobile transactions as a set of subtransactions. They introduce reporting and co-transactions. A reporting transaction can share its partial results, can execute concurrently and can commit independently. Co-transactions are like co-routines and are not executed concurrently. The model allows transactions to be executed on disconnection. It also supports unilateral commitment of subtransactions, compensating and non-compensating transactions. The author claims that the model minimizes wired as well as wireless communication cost. However, not all the operations are compensated [39], and compensation is costly in mobile computing.

Transaction models for mobile computing that perform updates at mobile computers have been developed in [3,39]. These efforts propose a new correctness criterion [39] that is weaker than serializability. They can cope more efficiently with the restrictions of mobile and wireless communications.

In [37,40,41], we look at a mobile transaction more as a concurrency control problem and provide database consistency. We incorporate a prewrite operation [42] before a write operation in a mobile transaction to improve data availability. A prewrite operation does not update the state of a data object but only makes visible the value that the data object will have after the commit of the transaction. Once a transaction has received all the values read and declares all the prewrites, it can pre-commit at the mobile host (i.e., computer connected to unreliable communication) and the remaining transaction's execution is shifted to the stationary host (i.e., computer connected to the reliable fixed network). Writes on a database, after pre-commit, take time and resources at stationary host and are therefore, delayed. This reduces network traffic congestion. A pre-committed transaction's prewrite values are made visible both at the mobile and stationary hosts before the final commit of the transaction. This increases data availability during frequent disconnections that are common in mobile computing. Since the expensive part of the transaction's execution is shifted to the stationary host, it reduces the computing expenses (e.g., battery, low bandwidth, memory etc.) at the mobile host. Since a pre-committed transaction does not abort, no undo recovery needs to be performed in our model. A mobile host can cache only prewrite values of the data objects, which will take less space, time, and energy and can be transmitted over low bandwidth.

A kangaroo transaction (KT) model was given in [43]. It incorporates the property that transactions in a mobile computing hop from a base station to another as the mobile unit moves. The mobility of the transaction model is captured by the use of split transaction [44]. A split transaction divides on going transactions into serializable subtransactions. An earlier created subtransaction is committed and the second subtransaction continues its execution. The mobile transaction splits when a hop occurs. The model captures the data behaviour of the mobile transaction using global and local transactions. The model also relies on compensating transaction in case a transaction aborts. The model in [37] has the option of either using nested transactions or split transactions. However, the save point or split point of a transaction is explicitly defined by the use of pre-commit. This feature of the model allows the split point to occur in any of the cells. Unlike the KT model, the earlier subtransaction after pre-commit can still continue its execution with the new subtransaction since their commit orders in the model [37] are based on the pre-commit point. Unlike the KT, the model in [37] does not need any compensatory transaction.

In [45], a basic architectural framework to support transaction management in multidatabase systems is proposed and discussed. A simple message and queuing facility is suggested which provides a common communication and data exchange protocol to effectively manage global transactions submitted by mobile workstations. The state of global transactions is modelled through the use of subqueues. The proposed strategy allows a mobile workstation to submit global transactions and then disconnected itself from the network to perform some other tasks thereby increasing processing parallelism and independence.

A transaction management model for the mobile multidatabase is presented in [46], called Toggle Transaction Management Technique. Here site transactions are allowed to commit independently and resources are released in timely manner. A toggle operation is used to minimize the ill-effects of the prolonged execution of long-lived transaction.

In most of the above papers, there is no comparative performance evaluation of models presented. We observe that there is a need to investigate the properties of mobility by means of experiments, which can impact most the transaction processing. Also, there is a need to evaluate various transaction processing algorithms with respect to performance, response time, throughput and may be a new paradigm like quality of service (QoS) in the transaction management in an area such as e-commerce.

4.1 Broadcast Disk and Transaction processing

In traditional client-server systems, data are delivered from servers to clients on demand. This form of data is called pull-based. Another interesting trend is push-based delivery in a wireless environment. In wireless computing, the stationary server machines are provided with a relative high bandwidth channel that supports broadcast delivery to all mobile clients located inside the cell. In a push-based delivery, a server repetitively broadcasts data to clients without a specific request. Clients monitor the broadcast and retrieve data items they need as they arrive on the broadcast channel. This is very important for a wide range of applications that involve dissemination of information to a large number of clients. Such applications include stock quotes, mailing lists, electronic newsletters, etc.

Broadcast in a mobile computing has a number of difficulties. How to predict and decide about the relevance of the data to be broadcast to clients? One way is that the clients may subscribe their interests to services [17]. The server also needs to decide about either sending the data periodically [48] or aperiodically. Mobile clients are also resource poor and the communication environment is asymmetric. The problem also is to maintain the consistency of broadcast data. The commercial system which supports the concept of broadcast data delivery has been proposed in [49]. Recently, broadcast has received considerable attention in the area of transaction processing in mobile computing environments.

Pitoura et al. [50] addresses the problem of ensuring consistency and currency of client read-only transactions when the values are being broadcast in the presence of updates at the server. The authors propose to broadcast additional control information in the form of invalidation reports, multiple versions per item and serializability information. They propose different methods that vary in complexity and volume of control information transmitted and subsequently differs in response times, degree of concurrency, and space and processing overheads. The proposed methods are combined with caching to improve query latency.

Authors in [51] exploit versions to increase concurrency of client read-only transactions in the presence of updates at the servers. Invalidation reports are used to ensure the currency of reads. They broadcast older versions along with new values. The approach is scalable as it is independent of the number of clients. Performance results show that the overhead of maintaining older versions is low and at the same time concurrency increases. On the same line, [52] presents an approach for concurrency control in broadcast environments. They propose a weaker notion of consistency, called update consistency, which still satisfy the mutual consistency of the data.

Transactions support in a mobile database environment with the use of a broadcast facility is reported in [53]. Mobile clients use broadcast data to verify if transactions are serializable. [54] presents an optimistic concurrency control protocol with update timestamps to support transactional cache consistency in a wireless mobile computing environment using broadcast. They implement the consistency check on accessed data and the commitment protocol in a truly distributed fashion as a part of a cache invalidation process with most of the burden of consistency check being downloaded to mobile clients. They achieve improved transaction throughput in comparison to [53] and it minimizes the wireless communication required for supporting mobile transactions.

5.   MOBILE DATABASE RESEARCH DIRECTIONS

We see the following as upcoming mobile database research directions:

Location-dependent Query Processing

We present ideas for processing queries that deal with location-dependent data [55]. Such queries we refer to as location-dependent. Location can be a subject of more complex aggregate queries, such as finding the number of hotels in the area you are passing or looking for a mobile doctor closest to your present location. Hence, the location information is a frequently changing piece of data. The objective is to get the right data at different locations for processing a given query. The results returned in response to such queries should satisfy the location constraints with respect to the point of query origin, where the results are received, etc. We propose to build additional capabilities into the existing database systems to handle location-dependent data and queries.

We present some examples to recognize the problems of accessing correct data when the point of contact changes. Data may represent SSN (Social Security Number) of a person, or maiden name, or sales tax of a city. In one representation, the mapping of the data value and the object it represents is not subjected to any location constraints. For example, the value of the SSN of a person remains the same no matter from which location it is accessed. This is not true in the case of sales tax data. The value of the sales tax depends on the place where a sales query is executed. For example, the sales tax value of West Lafayette is governed by a different set of criteria than the sales tax of Boston. We can therefore, identify the type of data whose value depends on the set of criteria established by the location and another type of data, which is not subject to the constraints of a location. There is a third type of data that is sensitive to the point of query. We illustrate this data with the following example. Consider a commuter who is traveling in a taxi and initiates a query on his laptop to find the nearby hotels in the area of its current location. The answer to this query depends on the location of the origin of the query. Since the commuter is moving he may receive the result at a different location. Thus, the query results should correspond to the location where the result is received or to the point of the origin of the query. The difference in these two correct answers to the query depends on the location and not on the hotel. The answer to the query "find the cheapest hotel" is not affected by the movement. The former depends on the location whereas later depends on the object characteristics. [55] discusses the data organization issues in location dependent query processing.

In [56], queries with location constraints are considered. They consider the query such as "find the nearest hotel from my current position". The main objective there is to minimize the communication cost to retrieve the necessary information to answer the query. The authors suggest greedy heuristics to solve the problem.

In the MOST project [57,58], the authors consider a database that represents information about moving objects and their location. They argue that existing DBMS's are not well equipped to handle continuously changing data, such as location of moving objects. They address the issue of location modelling by introducing the concepts of dynamic attribute (whose value keeps changing), spatial and temporal query languages and indexing dynamic attributes.

View Maintenance in Mobile Computing

Accessing on-line database from a mobile computer may be expensive due to limited uplink bandwidth, and also due to the fact that sending messages consumes considerable energy which is limited in the portable battery. These two problems can be solved by maintaining a materialised view i.e., storing the tuples of the view in the database at the mobile computer. This view will be updated as the on-line database changes using wireless data messages. This will also localise access, thus improving access time. Therefore, to better deal with the problem of disconnection, reliability, and to improve response time, the view should be materialised at the mobile computer. The view maintenance will involve location-dependent data, time-dependent data, and dynamic allocation of a materialised view in the fixed and mobile network. Another problem is dynamic allocation of a materialized view in the mobile network [30]. More work in this direction has been reported in [59]. Another issue concerns the divergence of the materialised view at the mobile computer from the on-line database. In other words, how closely should the materialised view reflect the on-line database. An approach given in [60] involves parameterising each read at the mobile computer with the amount of divergence from the latest version it can accept. Another approach given in [61] allows the user to specify triggers on the on-line database and the view is updated when the trigger is satisfied. Recently, this problem has also been discussed in a position paper [62] to emphasise the importance of data warehousing in view maintenance in a mobile computing environment.

There is a need to develop view maintenance algorithms in the case of data warehousing environment where relational data is broadcast. Similarly, there is a need to do change management in the web data where changes are broadcast periodically and mobile host should capture the data and make the cached web data consistent.

Workflows in Mobile Environment

Workflow management systems are growing due to their ability to improve the efficiency of an organization by streamlining and automating business processes. Workflow systems have to be integrated with the mobile computing environment [63] in order to co-ordinate a disconnected computer to enhance the system's resilience to failures. Specific issues that arise here include how the workflow models can co-ordinate tasks that are performed when mobile users work in disconnected mode and when they cross wireless boundaries. Also location sensitive activities might have to be scheduled to use an organization's resources effectively. Current workflow systems do not seem to have any provision to handle these requirements.

Authors in [63] discuss how disconnected workflow clients can be supported while preserving the correctness of the overall execution and allowing co-ordinated interactions between the different users. The required activities (i.e., applications and data) are downloaded onto the mobile computer before performing a planned disconnection. The activities are performed in a disconnected mode and the results are then uploaded after reconnection at which time exceptions that occurred during the disconnected mode of operations are also handled. When mobile users cross the borders of wireless cells, a hand-off (i.e., a mobile user's session is to migrate to a new information server) may need to be performed from the old workflow server to a new server. Consistency issues that arise when workflow instances migrate are to be handled carefully. For location sensitive routing of a mobile user's request, the modelling primitives should have a provision to specify geographic information in the workflow definition.

Mobile Web and E-commerce

There is a need to bring the web onto a mobile platform. Imagine a taxi that is equipped with a mobile computer and a passenger would like to browse web pages while waiting to reach their destinations. The limited bandwidth will be a bottleneck in such a scenario. Another interesting application may be e-commerce on the mobile web. All these applications are viable for the disconnected, highly unreliable, limited bandwidth and unsecured platform, but where the application demands reliability and security. Some efforts in this direction have appeared in [64], which is a WWW system designed to handle mobile users. It allows the documents to refer and react to the current location of clients. In [65], they address the issue of mobile web browsing through a multi-resolution transmission paradigm. The multi-resolution scheme allows various organizational units of a web document to be transferred and browsed according to the information contents. This will allow better utilization of the limited bandwidth. Similar effort has been reported in [66].

Mobile Data Security

Security is a prime concern in mobile databases due to nature of the communication medium. New risks caused by mobility of users and portability of computers can compromise confidentiality, integrity and availability-including accountability. In a mobile database environment, it may be a good idea to have data summarized [67], so that only metadata can be stored on mobile platform and more detailed data can be kept only on the mobile service station (MSS). The higher frequency of disconnection also requires a more powerful recovery model. Such situations offer attackers the possibility of masquerade as either a mobile host or MSS. This needs a more robust authentication service [68]. Another issue is to maintain the privacy of location data of mobile hosts. Ideally, only mobile user and home agent should have knowledge about a mobile host's current position and location data. All user identification information, including message origin and destination, has to be protected. In order to achieve anonymous communication, aliases can be used or communication can be channelled through a trusted third party. Furthermore, the identity of users may also need to be kept secret from other MSSs. Access-based control policies can be adapted to provide data security on a mobile platform.

6.   CONCLUSIONS

Mobility brings in a new dimension to the existing solutions to the problems in distributed databases. We have surveyed some of the problems and existing solutions in that direction. We have highlighted the merits and demerits of existing solutions. We have identified some of the upcoming research areas that, due to the nature and constraints of mobile computing environment, require rethinking. The upcoming mobile database research directions discussed here will be the centres of attractions among mobile database researchers in years to come.

REFERENCES

[1] B. Bruegge and B. Benninington, "Applications of Mobile Computing and Communications," *IEEE Personal Communications*, vol.3, no. 1, Feb. 1996.

[2] T. Imielinksi and B. R. Badrinath, "Wireless Mobile Computing: Challenges in Data Management," *Communications of ACM*, 37(10), October 1994.

[3] E. Pitoura and B. Bhargava, "Building Information Systems for Mobile Environments," *Proc. of 3rd International Conference on Information and Knowledge Management*, 1994, pp.371-378.

[4] E. Pitoura and B. Bhargava, "Maintaining Consistency of Data in Mobile Computing Environments," *Proc. of 15th International Conference on Distributed Computing Systems*, June,1995. Extended version appeared in *IEEE Transactions on Knowledge and Data Engineering*, 2000.

[5] J. Ioannidis and G.Q. Maquire, "The Design and Implementation of a Mobile Networking Architecture," *USENIX winter 1993 Technical Conference*, Jan. 1993.

[6] S. Mohan and R. Jain, "Two user location Strategies for PCS," *IEEE Personal Communications Magazine*, 1(1), 1st quarter, 1994.

[7] D.J. Goodman, "Trends in Cellular and Cordless Communications," IEEE *Communication Magazine*, June 1991.

[8] B. Awerbuch and D. Peleg, "Online Tracking of Mobile Users," *Journal of ACM,* vol. 42, no. 5, 1995, pp. 1021-1058.

[9] B. Badrinath, R. Imielinski and A. Virmani, "Locating Strategies for Personal Communication Networks," *Proc. of IEEE GLOBECOM workshop on Networking for Personal Communications Applications*, Dec. 1992.

[10] G. Cho and L.F. Marshall, "An efficient Location and Routing Scheme for Mobile Computing Environment," *IEEE Journal on selected Areas in Communications*, vol. 13, no. 5, June 1995.

[11] N. Shivakumar and J. Widom, "User Profile Replication for Faster Location Lookup in Mobile Environments," *Proc. of the 1st ACM International Conference on Mobile Computing and Networking (Mobicom '95)*, Oct. 1995, pp. 161-169.

[12] R.K. Ahuja, T.L. Magnanti and J.B. Orlin, *Network Flows*, Prentice Hall, Englewood Cliffs, New Jersey, 1993.

[13] R. Jain, "Reducing Traffic Impacts of PCS using Hierarchical User Location Databases," *Proc. of the IEEE International Conference on Communicatios*, 1996.

[14] M.V. Steen, F.J. Hauck, P. Homburg and A.S. Tanenbaum, "Locating Objects in Wide-area Systems," *IEEE Communication Magazine*, pp. 2-7, January 1998.

[15] E. Pitoura and I. Fudos, "An Efficient Hierarchical Scheme for Locating Highly Mobile Users," *ACM proceedings for International Conference on Information and Knowledge Management (CIKM)*, 1998.

[16] V. Anantharaman, M.L. Honig, U. Madhow and V.K. Wei, "Optimisation of a Database Hierarchy for Mobility Testing in a Personal Communication Network. Performance Evaluation," 20(13), May 1994.

[17] P. Krishna, N.H. Vaidya and D.K. Pradhan, "Static and Dynamic Location Management in Mobile Wireless Networks," *Journal of Computer Communications* (special issue on Mobile Computing), 19 (4), March 1996.

[18] S. Dolev and D.K. Pradhan, "Modified Tree Structure for Location Management in Mobile Environments," *Computer Communications*, vol. 19, 1997, pp.335-345.

[19] J.S.M. Ho and F. Akyildiz, "Dynamic Hierarchical Database Architecture for Location Management in OCS Networks," *IEEE Transactions on Networking*, vol. 5, no. 5, Oct. 1997.

[20] M. Satyanarayanan, "Mobile Information Access," *IEEE Personal Communications*, vol.3, No. 1, Feb. 1996.

[21] M. Satyanarayanan, J.J. Kistler, B. Kumar, et al., "Coda: A Highly Available File System for a Distributed Workstation Environment," *IEEE Transaction on Computers*, vol. 39, no. 4, April 1990.

[22] D. Barbara and T. Imielinski, "Sleepers and Workaholics: Caching Strategies in Mobile Environments," *VLDB Journal*, 1995.

[23] K. Wu, P.S. Yu and M. Chen, "Energy Efficient caching for Wireless Mobile Computing," *Proc. of the 12th International Conference on Data Engineering*, New Orleans, Feb. 1996.

[24] M.R. Ebling, "Evaluating and improving the Effectiveness of Caching for Availability," Ph.D. Thesis, Department of Computer Science, Carnegie Mellon University, 1997.

[25] J. Cai, K.L. Tan and B.C. Ooi, "On Incremental Cache Coherency Schemes in Mobile Computing Environment," *Proc. of IEEE International Conference on Data Engineering (ICDE)*, 1997.

[26] O. Wolfson and S. Jajodia, "Distributed Algorithms for Dynamic Replication of Data," *Proc. of the Symposium on Principles of Database Systems*, CA, 1992, pp. 149-163.

[27] B.R. Badrinath, and T. Imielinski, "Replication and Mobility," *Proc. of 2nd IEEE Workshop on the Management of Replicated Data*, Nov.1992, pp. 9 -12.

[28] K. Ravindran and K. Shah, "Casual Broadcasting and Consistency of Distributed Data," *Proc. of 14th International Conference on Distributed Computing Systems*, June 1994, pp. 40-47.

[29] M. Faiz and A. Zaslavsky, "Database Replica Management Strategies in Multidatabase Systems with Mobile Hosts," *Proc. of 6th International Hong Kong Computer Society Database Workshop*, 1995.

[30] Y. Huang, P. Sistla and O. Wolfson, "Data Replication for Mobile Computers," *Proc. of the ACM SIGMOD International Conference on Management of Data*, 1994.

[31] J. Gray, P. Helland, P. O'Neil and D. Shasha, "The Dangers of Replication and a Solution," *Proc. of ACM SIGMOD International Conference on Management of Data*, 1996, pp. 173-182.

[32] S. Wu and Y. Change, "An Active Replication Scheme for Mobile Data Management," *IEEE Proceedings of 6th DASFAA*, Taiwan, 1999.

[33] N. Barghouti and G. Kaiser, "Concurrency Control in Advanced Database Applications, " *ACM Computing Surveys*, vol. 23, no. 3, 1991, pp.269-317.

[34] K. Ramamritham and P.K. Chrysanthis, "A Taxonomy of Correctness Criterion in Database Applications," *Journal of Very Large Databases,* vol. 4, no. 1, Jan. 1996.

[35] G.D. Walborn and P.K. Chrysanthis, "Supporting Semantics-Based Transaction Processing in Mobile Database Applications," *Proc. of 14th IEEE Symposium on Reliable Distributed Systems*, Sept. 1995, pp.31-40.

[36] J. Kisler and M. Satyanarayanan, "Disconnected Operation in the Coda File System," *ACM Transactions on Computer Systems*, vol. 10, no. 1, 1992.

[37] S.K. Madria and B. Bhargava, "Improving Availability in Mobile Computing Using Prewrite Operations," *Distributed and Parallel Database Journal*, Sept. 2001.

[38] Q. Lu and M. Satyanaraynan, "Improving Data Consistency in Mobile Computing Using Isolation-Only Transactions," *Proc. of the fifth workshop on Hot Topics in Operating Systems*, Orcas Island, Washington, May 1995.

[39] P.K. Chrysanthis, "Transaction Processing in a Mobile Computing Environment," *Proc. of IEEE workshop on Advances in Parallel and Distributed Systems*, Oct.1993, pp.77-82.

[40] S.K. Madria and B. Bhargava, "A Transaction Model for Mobile Computing," *Proc. 2nd IEEE International Database and Engineering Application Symposium (IDEAS'98)*, Cardiff, U.K., 1998.

[41] S.K. Madria and B. Bhargava, "On the Correctness of a Transaction Model for Mobile Computing," *9th Intl. Conf. on Database and Expert System Applications (DEXA'98)*, Vienna, Austria, Aug. 1998, *Lecture Notes in Computer Science*, vol. 1460, Springer-Varleg.

[42] S.K. Madria and B. Bhargava, "System Defined Prewrites to Increase Concurrency in Databases," *Proc. of First East-European Symposium on Advances in Databases and Information Systems* (in co-operation with ACM-SIGMOD), St.-Petersburg, Sept. 1997.

[43] M. H. Eich, and A. Helal, "A Mobile Transaction Model That Captures Both Data and Movement Behavior." *ACM/Baltzer Journal on Special Topics on Mobile Networks and Applications*, 1997.

[44] C. Pu, G. Kaiser and Hutchinson, "Split-transactions for Open-ended Activities," *Proc. of the 14th VLDB Conference*, 1988.

[45] L.H. Yeo, and A. Zaslavsky, "Submission of Transactions from Mobile Workstations in a Co-operative Multidatabase Processing Environment," *Proc. of the 14th IEEE International Conference on Distributed Computing Systems (ICDCS'94)*, June 1994.

[46] R.A. Dirckze and L. Gruenwald, "A Toggle Transaction Management Technique for Mobile Multidatabases," *ACM Proc. of International Conference on Information and Knowledge Management (CIKM)*, 1998.

[47] D.B. Terry, D. Goldberg, D.A. Nichols and B.M. Oki, "Continuous Queries over Append-only Databases," *Proc. of the ACM-SIGMOD International Conference on Management of Data*, June 1992.

[48] S. Acharya, R. Alonso, M. Franklin and S. Zdonik, "Broadcast Disks: Data Management for Asymmetric Communication environments," *Proc. of the ACM SIGMOD Conference*, California, 1995.

[49] Vitria Technology Inc. , http://www.vitira.com.

[50] E. Pitoura and P.K. Chrysanthis, "Scalable Processing of Read-only Transactions in Broadcast Push," *Proc. of IEEE International Conference on Distributed Computing Systems*, 1999.

[51] E. Pitoura and P.K. Chrysanthis, "Exploiting Versions for Handling Updates in Broadcast Disks," *Proc. of VLDB*, 1999.

[52] J. Shanmugasundaram, A. Nithrakashyap, R. Sivasankaran and K. Ramamritham, " Efficient Concurrency Control for Broadcast Environments," *Proc. of the ACM SIGMOD Conference*, 1999.

[53] D. Barbara, "Certification Reports: Supporting Transactions in Wireless Systems," *Proc. of 17th International Conference on distributed Computing systems*, 1997, pp. 466-473.

[54] S. Lee, C. Hwang and H. Yu, "Supporting Transactional Cache Consistency in Mobile Database Systems," *Proc. of ACM International workshop on Data engineering for Wireless and Mobile Data Access*, 1999.

[55] S.K. Madria, B. Bhargava, E. Pitoura and V. Kumar, "Data Organization Issues in Location Dependent Query Processing in Mobile Computing Environment," *Proc. of 4th East-European Symposium on Advances in Databases and Information Systems* (in co-operation with ACM-SIGMOD), Prague, Czech Republic, 2000.

[56] T. Imielinksi and B.R. Badrinath, "Querying in Highly Distributed Environments," *Proc. of the 18th VLDB Conference*, 1992, pp. 41-52.

[57] O. Wolfson, X. Bo, C. Sam and L. Jiang, "Moving Objects Databases: Issues and Solutions," *Proc. of SSDBM*, 1998, pp. 111-122.

[58] O. Wolfson, A.P. Sistla, C. Sam and Y. Yelena, "Updating and Querying Databases that Track Mobile Units," *Distributed and Parallel Databases,* vol. 7, no. 3, 1999, pp.257-387.

[59] G. Dong and M. Mohania, "Algorithms for View Maintenance in Mobile Databases," in *proceedings of the First Australian Workshop on Mobile Computing and Databases and Applications* (*MCDA '96*), Melbourne Australia, 1996.

[60] Y. Huang, P. Sistla and O. Wolfson, "Divergence Caching in Client-Sever Architectures," *Proc. of the third International Conference on Parallel and Distributed Systems* (*PDIS*), Austin, TX, Sept. 1994, pp. 131-139.

[61] P. Sistla and O. Wolfson, "Temporal Conditions and Integrity Constraints in Active Database Systems," *Proc. of the ACM SIGMOD International Conference on Management of Data*, May 1995, pp. 269-280.

[62] I. Stanoi, D. Agarwal, A. El Abbadi, S.H. Phatak and B.R. Badrinath, "Data Warehousing Alternatives for Mobile Environments," *Proc. of ACM International Workshop on Data Engineering for Wireless and Mobile Access*, Seattle, Washington, 1999.

[63] G. Alonso, R. Gunthor, M. Kamath, D. Agrawal, A. El Abbadi and C. Mohan, "Exotica/FMDC: Handling Disconnected Clients in a Workflow Management Systems," *Proc. of 3rd International Conference on Co-operative Information Systems*, May 1995.

[64] G.M. Voelker and B.N. Bershad, "Mobisaic: An Information Management System for Mobile Wireless Computing Environment," T. Imienlinski and H. Korth, editors, *Mobile Computing*, Kluwer Academic Publishers, 1996, pp. 375-395.

[65] M. T. Stanley, Y. H. Va Leong, D. McLeod and A. Si, "On Multi-Resolution Document Transmission in Mobile Web," *SIGMOD Record*, vol. 28, no. 3, 1999, pp.37-42.

[66] M. Gaedke, M. Beigl, H. Gellersen and C. Segor, "Web Content Delivery to Heterogeneous Mobile Platforms," *ER workshops Proc. as Lecture Notes in Computer Science*, vol. 1552, Springer, 1998.

[67] S.K. Madria, M. Mohania and J. Roddick, "A Query Processing Model for Mobile Computing using Concept Hierarchies and Summary Databases," *Proc. of 5th Intl. Conference on Foundation for Data Organization* (*FODO'98*), Japan, Nov. 1998.

[68] B. Bhargava, S. Kamisetty and S.K. Madria, "Fault Tolerant Authentication in Mobile Computing," *Sp. Session, New Paradigms in Computer Security, at International Conference on Internet Computing*, (*IC'2000*), Las Vegas, USA, pp. 395-402.

Chapter 9

A LOCAL/GLOBAL STRATEGY BASED ON SIGNAL STRENGTH FOR MESSAGE ROUTING IN WIRELESS MOBILE AD-HOC NETWORKS

Ting-Chung Tien and Shambhu J. Upadhyaya
State University of New York, Buffalo

Abstract: Route switching is an important issue to connection-oriented communication that needs to maintain a route for a certain period of time. This chapter presents a new approach to ease the route switching problems in a network with a high rate of unit migration. By monitoring the signal strength of messages, a unit in a route that receives an incoming message can detect possible route fluctuations locally. As the average signal strength declines into a dangerous level, the unit that receives the message will send an advance-warning message to the route source unit. If the source unit can find more stable routes locally, it will adapt a substitute route and will complete the process of adaptation before the breakdown of the original route. If the route source unit cannot adapt a new route locally, the source unit will be forced to search for a new route by considering the entire network. The evaluation conducted quantifies the benefits of the new approach.

Keywords: Ad-hoc Networks, Mobile connectivity, Real-time content transmission, Routing, Wireless/terrestrial interfaces

1. INTRODUCTION

In a wireless environment, battery power and frequency channels limit the message routing among mobile units. Traditional wired networks normally do not have such limitations, and their units also have more powerful network computing capabilities than that of mobile units. Besides these limited resources, the mobility characteristics of the mobile units are quite different from the computing units of the wired networks. In the wired networks, the movement of the units and the network topology changes are rare. But it often happens in a distributed mobile wireless computer network, especially for the wireless mobile Ad-hoc network. The main challenge for a wireless mobile Ad-hoc network comes from its inherent characteristic of high uncertainty. Therefore, to deal with the routing problems within a distributed wireless mobile computer network, one should resort to the concepts that are different from that of the traditional wired data computing network architectures [1,2,3,4,5].

Most of the wireless data computing networks can be classified into two types of architectures: 1) base station supported, such as the wireless Ether LAN, and 2) no base station supported, such as the wireless mobile Ad-hoc networks. For the wireless LAN, the base stations generally support any network computing needed of its subordinate units for their local network access and services. The wireless mobile Ad-hoc network must depend on its subordinate units to do their own network computing and all these units in this network act as dynamic distributed routers [6,7,8].

1.1   The Network Routing Algorithms

The main purpose of the various routing algorithms is to guarantee an error free message transmission from source to destination and maintain its correctness. There are three major algorithms utilized in today's traditional wired networks [9], namely, the distance vector routing [10], the link state routing [11], and the source routing [9]. Basically, these routing algorithms try to take the shortest path approach to route data in a network [12]. However, there are several inherent weaknesses with these algorithms.

The link state routing algorithm uses a centralized approach where heavy route computations and the broadcast of massive maintenance information increase the load on the units and the consumption of extra power. Also, the maintenance information should be broadcast in relatively short duration to respond to the dynamic environments. The massive broadcasting may degrade the data transmission traffic. The route computation time could be worse since it is proportional to the cube of the number of the units in the network [13].

The distance vector routing is based on the distributed Bellman-Ford algorithm. The inherent weaknesses of this algorithm are two-fold: the periodical broadcasting may cause network congestion and consume extra power, and the other drawback is the oscillation problem. The oscillation problem occurs when a selected route increases its routing distance, which makes the selection of another route next time. However, the second route increases the routing distance that makes the first route more desirable at the subsequent routing. Eventually, the routing will be oscillatory between the two routes. This problem is caused by the feedback effect between link lengths and routing updates [9]. For a highly dynamic environment, the previous broadcasting of link information may be unable to reflect the newest changes of links in time.

The source routing algorithm has been used in today's bridged local area networks. This routing lets a source unit determine a complete sequence of units through which to forward packets to the destination, and explicitly lists this route in the header of the forwarding packets. This routing algorithm tries to discover a route from the source to the destination by broadcasting (or by flooding) an exploratory message through the extension of the network. The destination sends a route-confirmed message back to the source unit through each possible loop-free route. From the many possible routes, the one that satisfies the shortest path criterion is selected. The addresses of all the units in the route are included in the header of each message communicated between the source-destination pair to establish a virtual circuit. All the data messages follow the same route. The inherent weakness of this algorithm is that in order to update the forwarding database or to establish new connections between the source-destination pairs, all the units have to periodically broadcast messages throughout the network and these actions may cause the problems of link congestion and power consumption.

In 1996, Johnson and Maltz proposed the dynamic source routing (DSR) algorithm [14]. This routing algorithm is based on the source routing algorithm with an improvement for the dynamic environment. It is explicitly designed for use in the wireless environment of an Ad-hoc network. Because it does not periodically broadcast the routing advertisement, it greatly reduces the network bandwidth overhead and the battery power consumption. In 1997, a new signal stability-based adaptive routing (SSA) algorithm [15] tried to find the most stable route by analyzing the signal strength. This SSA algorithm has further improved the communication quality for the message routing in a wireless Ad-hoc mobile network. Besides these above two algorithms, there have been many attempts from different researchers trying to solve some basic problems for such networks [16,17,18,19,20].

1.2   Problems with Message Routing in Wireless Ad-hoc Mobile Networks

As an intermediate route unit migrates from a route, message transmission could suffer a route failure. Searching for a substitute route after the initial route encounters a route failure will postpone all the following message transmissions, and decrease the throughput of the entire

Ad-hoc network. It even could lead to the loss of some precious data. As the migration rate of a unit increases, the data throughput of Ad-hoc networks gets worse. The process of route discovery is both making network throughput decline and delaying the transmission of time-concerned messages. In the worst case, the entire network communication would be degraded. However, it is important to allow the mobile units to move around in an Ad-hoc network, and in the meantime maintain communication services at an acceptable level. Thus, to deal with the message routing in an Ad-hoc network with high unit migration rate, it is essential to devise new routing schemes.

In the algorithms of the DSR and the SSR, the source-destination pair units utilize a route until all message exchanges have ended or a route-failure has occurred. When a message routing encounters a link failure, such as the network partition or a route unit breakdown, the route unit encountering the message routing failure sends a routing-error message (REM) to the source unit to indicate the routing-error situation. The REM points out where and when the routing-error occurred. The source unit then launches a process of route discovery for the route-failure under two conditions: 1) the source unit has got the routing-error message, and 2) the routing-error message cannot reach the source unit, and the waiting for an acknowledgment message until a time-out expires is over and re-sending the same message to the destination until a specified number of tries also exceeded. The new route searching procedure is the same as that of the initial route built between the source-destination pair. During this period of the route reconstruction, the throughput of the source-destination pair units decreases greatly. In an Ad-hoc network with high mobility, the whole network performance may be degraded. This means that the higher the migration rate of mobile units, the lower the throughput of the network. In this chapter, we propose a new scheme to improve the communication efficiency for Ad-hoc networks under highly dynamic conditions. In this scheme, a threshold called warning level is devised to mark a deteriorating communication link by analyzing the strength of the incoming signal. By quantifying deteriorating links for each available route, it is possible to find a relatively stable route to route messages during the global adaptation phase. To reduce the communication latency due to any potential link failure, a local adaptation algorithm is proposed that finds a replacement route locally without the need to launch a new global route discovery process. Meanwhile, the original route can keep routing messages without interruption. In Section 2, the basic principle of the new routing scheme is given. Section 3 describes the global route adaptation strategy and Section 4 discusses the development of the new local route adaptation algorithm. Evaluation, results and a discussion appear in Section 5.

2. A NEW ROUTING SCHEME BASED ON SIGNAL STRENGTH

2.1 Basic Idea

We employ an advance warning mechanism (AWM) to offer an early warning service for message exchanges between wireless mobile units. This mechanism monitors each route link situation and responds to any possible route fluctuations before the actual link failure due to the unstable signals over free space. The concept of utilizing the signal strength to design an algorithm for the message routing in the wireless networks is not a new idea [15]; it has been practiced in wireless communication networks such as the cell phone and the PCS networks. However, the design of a routing algorithm with a local adaptation capability by combining signal strength monitoring with an advance warning mechanism to offer and maintain a stable communication service is a new attempt for message routing in a wireless Ad-hoc mobile network. The AWM acknowledges possible route fluctuations by monitoring the signal strength from its neighborhood, including the previous hop and the next hop along a route. If the signal strength of a received message reaches a strength level called the warning point, the AWM will result in a warning message, which will be sent to the previous hops of the incoming link. The previous hops will try to adapt a substitute route locally. If this local adaptation fails, it will go into the phase of global adaptation. This should give the source unit enough time to make optimal decisions, whether it should still utilize the initial route or launch a route request procedure to find a new one, based on the information status before the route-failure actually occurs. We assume that each unit along a route has an AWM and the capability to listen to its neighborhood, so that the reliability of the message routing and the network throughput can be ensured. Unlike other routing schemes, our scheme could respond more quickly to a route fluctuation.

2.2 Signal Strength Determination

Considering the fact that the stability of the signal transmission strength cannot be guaranteed in free space, it is desirable to introduce a mechanism to ensure the quality of the message exchanges between units. In the wired networks, the signal transmission quality over a conductor wireline can be guaranteed over a limited distance (for example, the 10Base5 is 500m for a segment, and 2500m for the total length) [21]. The wireless networks transmit signal over free space, and the average energy of a message signal (s_{av}) decreases by the inverse square of the signal transmission range ($s_{av} \propto$ watts/meter^2) [22]. If a unit wants to cover more regions, the power needed to transmit a signal must be increased. Because the relative distance is a factor of the signal strength between the sender and the detector, we set up a mechanism to acknowledge the change of a relative distance and the units' movement by detecting the change in the signal strength. The fluctuation of the signal strength could also come from other factors besides the units' movement, such as the low battery power of units and the geological effect. Because they also cause the signal instability, the early warning concept still can be utilized in those cases. The signal strength determination is described in Figure 1.

In Figure 1, the horizontal axis is the separation between two mobile units. Distance d_2 is the maximum range for effective message communication. Beyond this point, the link will be highly unstable that might lead to communication interruption. Distance d_1 is the experimentally decided warning point. The difference between d_1 and d_2 is the warning range.

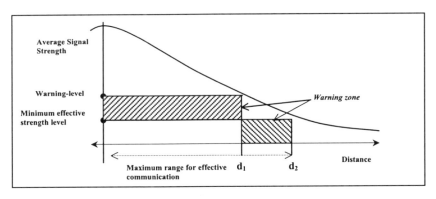

Figure 1. Signal strength waveform

3.   GLOBAL ADAPTATION STRATEGY

3.1   Route Discovery and Data Routing

Our route discovery scheme is primarily based on the source routing algorithm [9] and the DSR [14]. This type of data routing is also called route-on-demand. Like the DSR and the SSA schemes, a source unit will broadcast a route request message (RQM) to find a route to the desired target unit if it does not have a usable cached route in its own routing table. The RQM will record each unit's address it passes until reaching the destination. Each unit in the network maintains a routing table that keeps the lists of routes and the active neighborhood units. In this routing table, there is a warning mark number associated with each route and a warning mark associated with each neighborhood link. A counter is used to record how many warning links the RQM has encountered, and it starts with 0. Every unit that receives a RQM will analyze the signal strength of this message, and determine whether the incoming message has reached the warning point or not. If the incoming link declines to the warning point, a warning will be marked on this link of the routing table, and the warning number counter on this RQM will be increased by 1. A unit could receive several RQMs from different routes. It can first rebroadcast the first incoming RQM anyway, and rebroadcast later one with fewer warning marks if this unit is not the destination. The units can also limit the maximum number of the allowed warning marks (WM) of a RQM. The receiver units will discard every RQM with the higher WM number. The destination unit could receive several RQMs from different routes but will choose the first incoming RQM with 0 WM number. Otherwise, one RQM with the least WM number will be the choice. By selecting the least WM number at each unit, the destination can find the most stable route. The destination unit will send the source unit a route reply message (RRM) to respond to the selected route. There are several approaches for the destination unit to send the RRM to the source unit of a RQM [14,15].

1) If the destination unit has an entry for the origin in its route cache, it may use this source route to send a RRM.

2) The destination unit can use the reverse route to send back RRM. This scheme requires the link between each pair of units working equally well in both directions.

3) The destination unit can piggyback the RRM on a RQM targeted at the original unit, and try to find a route from destination to the original unit by broadcasting an RQM with piggybacked RRM.

After the source unit receives the RRM, it could start to send the first data message through the route in the route record (RR) from an RRM. Each unit recorded in the RR will register this route in its own routing table and retransmit the first data message to the next intermediate unit. As the destination unit receives this data message, a route is set up completely. If approach 2 is adapted for replying an RRM, there is another possible way to build up a message route. During an RRM send-back to the original unit through this reversed route, each intermediate unit can set up this route simultaneously. Thus, as an RRM reaches the original unit, a communication route has been set up.

In the traditional source routing algorithm the addresses of all units in the route are included in the header of the data routing protocols. This will reduce both data message and network bandwidth efficiency. Our new scheme also utilizes routing tables but does not include all units' addresses in the header of the data messages. The routing table maintains a list of route records, and each route record is associated with a routing number and a warning mark counter. Each route record keeps the address of every route unit. The header of the data message (except the first data message if the third approach of RRM replying is adapted) includes both addresses of the source and the destination, a routing number, and a sequence number. Different messages transmitted from the source unit through the same route will have the same routing number but different sequence numbers. A unit receiving an incoming message will check the destination address and the sequence number to see if it is the destination and whether or not the same message has been received before.

By recognizing the sequence number, the units can prevent the possible looping situation. If and only if a unit receives a data message with a sequence number never been received before (which shows the unit is not the specified destination but does belong to the current route) it will retransmit this message to the next unit, until reaching the destination. If a unit receives a message for which it is not the desired destination and also does not belong to the current route, it will just discard the received message.

3.2   Route Maintenance and Routing Error

Consider a scenario where only a part of the network is static. From the perspective of the static topology, a route can be cached for later use again. But for the partially dynamic one, a cached route could be useless in a short period, especially for those units with a high migration rate. If a unit tries to send messages through a cached route obtained from earlier route discoveries and part of this route no longer exists, this message will encounter routing problems. If the units always process the route discovery before sending data and ignore the possible usage of the cached route, it would consume extra network resources.

Most proposed Ad-hoc network algorithms adopt the method of distributed network computing, which means that each unit will keep a local routing table. The local routing table reflects the network topology and it should be updated as the network topology changes. There are two major approaches to maintaining the local routing table, one active in which a unit broadcasts the link-checking messages or sends beacons to poll its neighbor units at fixed periods and the other passive in which a unit listens to its neighborhood and updates the routing table according to the result of eavesdropping [14,15]. Our scheme adopts a combination to get the advantages from both the passive and the active approaches. All units need not keep polling their neighborhood at every fixed time slot. It will be done on a need only basis. Based on this, every unit in an Ad-hoc network listens to its neighborhood; a mechanism is set up to figure out a proper time to launch a link-checking message (LCM) to a specified neighbor unit that has been silent too long. The mechanism of the LCM will turn back a timer to 0, which counts up a time-out for a silent neighbor unit and it starts from 0 after getting a signal from this specified unit, no matter whether it comes from the response of the LCM or eavesdrops from a neighbor's communication. The units will analyze every incoming message signal and update their local routing tables. If a neighbor unit no longer exists, it will be discarded from the local routing table of a LCM sender. Also, the signal strength of every message will be analyzed; the units will assign a warning mark on those incoming links that decline to the warning-range. If a neighbor unit is not silent within the maximum allowed time, then polling will not be necessary. Therefore, not only the battery power for the LCM can be saved but also the network bandwidth.

The routing failures in Ad-hoc networks might be caused by the migration of the route units, the unstable signal leading to interruptions between units, or the route unit failure itself. If a source unit continues to forward messages through the initial route and does not adapt the topology change when a route unit migrates away from a route, it might encounter a routing failure. But if the migrated unit is still in the route, it can possibly keep using the initial route by shortening the route [14]. If the instability factors come from the geographic topology, it could be a permanent effect. It would lead to the faulty data message transmissions that may require the buffered messages stored at the sender to be re-transmitted. Using the digital signal transmission and the advanced signal coding technique can improve the quality of the message transmission. The digital signal transmission is more robust than that of the analog [3,23,24]. The route unit failures, which can occur due to the malfunction or the power-off of this unit, will interrupt the message routing and cause the routing failures. The power-off of a unit can be foreseen by sending a notification to other units. The malfunction one is more difficult to deal with, because it can be hardly foreseen.

The problem with the earlier routing algorithms is explained with an illustration. In the previously proposed algorithms, the source-destination pair can exchange messages through a route till their communication is finished or until encountering a routing failure. If a route failure occurs during a communication and a route unit recognizes this, it sends a routing error message (REM) to the source unit. The source unit re-launches a new route discovery if it has received this message. To make sure that this route has failed, the unit encountering the routing problem has to wait until a preset time-out expires at the absence of any acknowledgment from the receiver. The sender may try the message transmission several times until a maximum number is reached, depending on the preset control information in the header of the data message [14,15]. The waiting delay plus the latency to produce an error message, the latency to send an error message back to the source unit, and the latency to re-launch a route discovery to search a substitute will induce an apparent delay in the communication. The cost for such message routing delay may sometimes become intolerable.

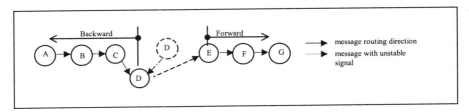

Figure 2. A simple migration scenario

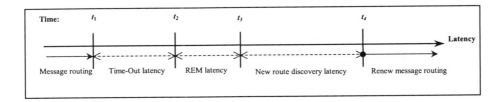

Figure 3. Time scheduler of renewing the message routing as a routing error occurs

Suppose that unit A is communicating with unit G through a group of units B, C, D, E, and F as shown in Figure 2. Assume that unit D is migrating away from this original route, and the separation between the units D and E is about to be larger than d, the maximum distance that D can communicate effectively with E. Then, this communication will encounter a message routing failure soon. For the existing hop-by-hop acknowledgment scheme, unit D will wait for the acknowledgment back from unit E to confirm the success of data transmission until a time-out expires. After the time-out, the unit either transmits the cached message once again or sends a route error message back to source A to indicate where and when the routing error occurred. If unit A has received this routing error message, it will launch a route discovery to look for a new route. As the destination unit *G* receives a RQM from *A*, an RRM with a newfound route will be sent back to unit *A*. When unit *A* receives the RRM from unit *G*, a new route has been found. Unit *A* will switch communication from the initial route to this alternate route. We will now analyze the total delay encountered to renew the communication between units *A* and *G*.

Following the example, assume that the route failed at the link between the units *D* and *E*. First, consider the case where this communication adopts the connection-oriented service and the hop-by-hop acknowledgment scheme. Unit *D* has just finished a message transmission to unit *E* at time t_1, and started a count-up counter from 0 to wait-for-an-acknowledgment from *E*. See Figure 3. The sender *D* will wait for a desired acknowledgment until the time-out at time t_2. If no acknowledgment is received from *E*, *D* either sends this cached message once again or starts to send a REM to the source unit *A*. Whether *D* re-sends the cached message or not depends on a preset number for a re-send times on the data message header. At time t_3, unit A receives the REM and re-launches a route discovery to look for a new route to unit *G*. By receiving the RRM from the destination *G*, unit *A* recognizes and registers that newfound route. Starting from time t_4, unit *A* renews the communication with unit *G*. The total time needed to renew a communication is $t_4 - t_1$. During this period, not only the throughput between the units *A* and *G* declines, but also the throughput of the entire Ad-hoc network faces a decline, even by using piggyback to improve the message efficiency. For the end-to-end acknowledgment approach, a time-out measurement is used for any possible routing failure. When the count-up timer reaches the limit of the time-out and still has not yet received an acknowledgment from the destination unit *G*, the source unit *A* will try to send the cached message once again or initiate a route discovery to find a new route. After unit *A* finds a new route by receiving the reply message sent from unit *G*, it renews the message routing to *G*. The timing for this approach is similar to that of the hop-by-hop scheme. The traditional algorithms for the route failure are slow to respond to the changes of the network topology. In the worse case that the change rate of the topology is faster than the updating rate for the new topology, the network will be impacted by the overhead of the route discovery. To avoid this situation, a mechanism is usually set up to limit the number of the tries and the interval to initiate the route query message. However, the earlier algorithms still lack enough capability to adapt to a dynamic network with high migration rate of mobile units.

4.   LOCAL ROUTE ADAPTATION STRATEGY

Before delving into the local routing adaptation algorithm based on the signal strength, it is necessary to analyze the route units' migration behaviors and their relative positions (RP). This is because these migration behaviors will affect the kinds of message transmission between units, and our local adaptation solutions will be based on these basic behavior models. There are three simple models to represent the basic behaviors caused by the migration of a route unit, and these basic models consist of all possible route units' migration behaviors in an Ad-hoc network. They are direction sensitive for the message routing, and for the sake of convenience, the message transmitting direction in the following models is from the left to the right. Each model is described in the following with a figure to show the relationship between the migrating unit and its neighbor units. In each figure, the *n*th unit always represents the leftmost unit that detects the migration movement of an intermediate unit.

4.1   Migration Models

Case 1: Migration Affecting Outgoing Link A route unit's migration movement affects its outgoing link transmission to the next hop but does not affect its incoming links. In this case, the next hop can detect that its incoming link is entering into the warning level. The picture on the left of Figure 4 shows that the (*n-1*)th route unit is migrating away from the *n*th unit, and the *n*th unit detects that the link between units of (*n-1*) and *n* is stepping into warning level.

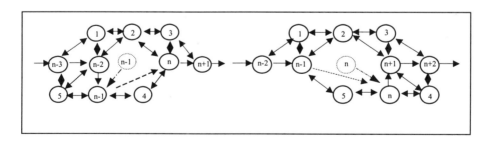

Figure 4. Migration affecting outgoing link (left), and migration affecting incoming link (right)

Case 2: Migration Affecting Incoming Link A route unit's migration movement affects its incoming link from a previous hop but does not affect its outgoing links to the next hops. This migrating route unit can detect that the incoming link is entering into the warning level. The picture on the right of Figure 4 shows that the *n*th route unit is migrating away from the (*n-1*)th unit, and the *n*th unit detects that the incoming link from the unit (*n-1*) is entering into the warning level.

Case 3: Migration Affecting both Incoming and Outgoing Links A route unit's migration movement affects both the incoming and the outgoing links simultaneously as shown in Figure 5. The (*n-1*)th route unit is migrating away from both units (*n-2*) and *n*. It can detect that the incoming link from unit (*n-2*) is entering into the warning level. Meanwhile, the *n*th unit also detects that the incoming link from unit (*n-1*) is entering into the warning level.

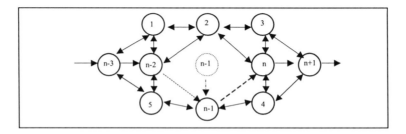

Figure 5. Migration affecting both incoming and outgoing links

From these models, it is seen that the links affected by a migrating unit are the link of (*n-2*) to (*n-1*) and the link of (*n-1*) to *n*, and the units that could be potentially involved in this event of migration are the units (*n-2*), (*n-1*), and *n*. The least number of units that could be affected is 2 for case 1 and case 2, or 3 for case 3. Therefore, the (*n-1*)th unit and the (*n-2*)th unit will be the prime units to make the decision for the local route switching. It is important to recognize every unit's relative position (RP) in a route. In order to recognize the position of a link in a route that encounters a warning condition, the list of cached routes in each unit's routing table arranges all route units' addresses by sequence ordering. The source unit is assigned the most significant position and the destination unit is at the least significant position. By this position ordering, each unit can recognize its relative position among the route units. From this position relationship, a unit can acknowledge itself belonging to the forward part or the backward part at each link along a route as shown in Figure 2.

4.2   Development of the Algorithm

As a receiver detects that an incoming link is at the warning level, it will do two things: 1) send a local warning message (LWM) backward to the sender that can be done either by locally broadcasting or by affixing a link warning (LW) to an acknowledgement, 2) increase the warning number by 1 at the header of the transmitted message. Thus, the destination unit gets informed that some link in this route is under warning. By checking the warning number at the header of the incoming message, the destination unit can monitor the conditions of all the links of a route.

4.2.1   Rules for Local Broadcasting

For the local broadcasting, the LWM transmission follows certain rules and regulations such that the set of available potential substitutes at a local region is small. These regulations are:
1) The LWM contains addresses of both the sender-end and the receiver-end, and the addresses of the route units.
2) Any forward unit with the RP > n+1 that receives a LWM just discards it to avoid unnecessary overhead in the network.
3) The LWM has a maximum hopping number.
4) The sender-end unit of a warning link will discard the LWM received directly from the receiver-end unit.
5) If any backward unit with the RP < n-2 receives a LWM without affixing any potential substitute candidate, it just discards this message.
6) The action of re-broadcasting is limited to a specific number of tries.
7) After broadcasting a LWM to its neighborhood, the nth unit will listen to its neighbor to confirm this message broadcasting. If no response, it can either rebroadcast this LWM or just stop this local broadcasting.

The rules 2 and 3 limit the potential substitutes at the local regions of the network. The rules 4, 5, and 6 are to reduce unnecessary transmission overheads. Rule 7 is to make sure that any possible existing neighbor units of the receiver-end unit of a warning link will receive the LWM. According to the results of the LWM received at the backward units, there are two possibilities for the local route discovery that depends on whether all links are capable of duplex communication, and there are two different approaches to deal with them.

4.2.2 Available Local Substitute Found by the LWM

The LWM message includes the entire route units' addresses, which allows the neighborhood of the sender-receiver pair to check if their routing tables have a potential candidate for such a local substitute route. A potential candidate should conjoin two intermediate route units of that initial route, and should not let the warning link still to be an intermediate part of this new route. By presetting the number for an acceptable local substitute, the potential candidates can be limited. To decide which substitute is adopted from several possible candidates, certain rules and regulations are defined. This will avoid any potential route-found messages competition.

1) If a neighbor unit has more than one potential local substitute at its routing table, it should choose the one with the least number of replacement units.

2) If a neighbor unit receives a LWM affixed with a candidate with the replacement units less than or equal to its own potential candidates at the routing table, this unit should re-transmit this LWM.

3) If a neighbor unit receives a LWM affixed with a candidate with a larger number of the replacement units than its own potential candidates, it should replace that by its own candidate to retransmit.

4) In order to limit the number of the potential candidates reaching a unit with the RP $\leq$ n-1, the first incoming candidate with the replacement units less than or equal to that of the initial route will be chosen as the substitute.

5) If several backward route units, RP $\leq$ n-1, have chosen their own candidates, the unit with the smallest RP and with the hopping number less than or equal to that of the initial route will be the one to do the local route-switching.

4.2.3 No Available Local Substitute Route Found by the LWM

If there is no available local substitute found with the LWM, the (n-1)th unit needs to broadcast a local route query message (LRQM), which is a RQM with a limited maximum hopping number (MHN), to the receiver unit. The transmission of the LRQM follows the regulations listed below.

1) Addresses of both the sender-end and the receiver-end, and the whole addresses of the route units are listed in the LRQM.

2) The LRQM limits the acceptable number of the replacement units for a potential substitute candidate.

3) The LRQM has a MHN that limits the number of units this message could pass to. Any unit that receives a LRQM with zero MHN should just discard it.

4) If a substitute candidate has been found before receiving a LRQM, the neighbor units should abandon the process of retransmission of the LRQM.
5) If the unit receiving a LRQM does not own any potential substitute, this unit should insert its own address in the route record, count down the limited hopping number by one, and retransmit this message.
6) Neighbor units receiving the LRQMs from a warning link should just discard it.
7) The units of the same initial route should discard the LRQM sent from the sender-end unit of a warning link.
8) If a unit that received a LRQM owns an acceptable potential candidate for this local substitute, it should send back its proposed candidate to the sender-end unit or any backward unit of this initial route according to its proposed candidate and stop retransmitting the LRQM.

The rules 2 and 3 limit the number of substitutes at the local region of the network. Rule 4 ensures avoidance of any unnecessary overhead. Rule 6 ensures that the receiver will not adapt a warning link in a substitute candidate. Rule 7 ensures that only the neighbor units of the sender-end of a warning link will receive the LRQM. Rule 8 ensures that there is no more transmission of a LRQM as a potential candidate and this substitute candidate is transmitted immediately back to some conjoint unit of this initial route.

The processing to adopt a proposed candidate from several possible potential substitutes at a neighbor unit and a conjoint route unit is the same as that of the previous approach. When a neighbor unit or a unit at the forward part adopts a local substitute, it will send a local route reply message (LRRM) immediately back to the backward unit, which will include a record of a proposed substitute candidate. As to set up a new local route for both cases, the record of this local substitute could be piggybacked to the data message and sent to the destination through this newfound route. Like the setup processing for the initial route, a substitute route is built by enlisting the addresses of the new intermediate units at the local routing tables of the units. After a new local substitute is found and set up, all the remaining message routing will be switched to this substitute.

However, there still has the possibility that the local adaptation is not successful, especially for the sparse network topologies. The local adaptation algorithm is limited to the partial networks, so the scheme of the global adaptation should be used if the local adaptation fails. To avoid unnecessary waiting, a time-out mechanism is set up at the destination unit. The duration of this time-out mechanism should depend on the real operating environments of the network, and it should be a statistical value obtained from experiments. As far as the waiting time-out is concerned, if the warning link still exists in a route and still maintains a warning mark, the destination unit should send a message to the source unit to indicate the link unstable situation. Eventually, the problem of route switching will go into the global adaptation phase.

4.3   Illustration

To see how to apply these regulations, we consider a simple example by using case 1 of the migration model. In case 1, as unit n detects that the link of $(n-1)$ to n is at the warning level, it broadcasts a LWM with the maximum hopping number equal to 2 (MHN = 2) to its neighborhood. The units $(n-1)$ and $(n+2)$ will discard this message, but unit $(n+1)$ will retransmit it. The neighbor units n, 2, 3, 4, and $(n+1)$ will check the route record that came with this LWM. Unit 3 will discard this LWM from unit $(n+1)$ because it has received this once from the nth unit. Unit 3 also will discard and refrain from retransmitting this LWM affixed with unit 2's proposed candidate, for unit 2 has smaller hopping number for the proposed candidate than its own proposed candidate. Unit 2 will discard and stop re-transmitting this message affixed with unit 3's proposed candidate, for unit 3 has the larger hopping number of proposed candidate, '-- $(n-2)$ -2-3-n --', than that of its proposed candidate. Unit 2 will retransmit the LWM from the nth unit and affix its own proposed candidate with this message, '-- $(n-2)$-2-n --'. Unit 4 will discard and stop re-transmitting the LWM affixed with unit 2's proposed candidate, for its proposed candidate, '-- $(n-2)$-4-2-n --', has the larger hopping number than that of unit 2. As the LWM affixed with unit 2's proposed candidate reaches at the route unit $(n-2)$, unit $(n-2)$ will immediately select this substitute to be its proposed candidate, for it has the same hopping number as that of the initial route.

Unit 4 will affix its proposed candidate, '-- $(n-2)$-$(n-1)$-4-n --', with this LWM from unit n and retransmit it. The route unit $(n-1)$ will accept unit 4's candidate to be its proposed candidate, for this substitute is the only one reaching at unit $(n-1)$. Then, both units $(n-2)$ and $(n-1)$ have their own proposed substitute candidates. Because the MHN is limited to 2, the possible units that received the LWM are the units $(n-2)$ and $(n-1)$ in this example. And as unit $(n-2)$'s RP and the hopping number of its proposal are smaller than that of unit $(n-1)$, this candidate of '-- $(n-2)$-2-n --' will be the local substitute automatically. Route unit $(n-2)$ will now switch the data messages to this new substitute route. The other models can also utilize the same process to find a local substitute for the initial route.

5.   EVALUATION, RESULTS AND DISCUSSION

In this section, we evaluate our local/global strategy based on the signal strength, by calculating the latency needed for the route switching when a routing failure occurs. First, we analyze the latency for both the global and local adaptation algorithms without utilizing the advance warning mechanism. Then we will add the advance warning mechanism to the local adaptation scheme to see how much improvement can be obtained. This evaluation demonstrates that our proposal has the potential to handle the route switching seamlessly for communication services at a mobile wireless Ad-hoc network that has the characteristic of high migration rate of mobile units.

We derive a simple function to estimate the latency for route switching for the global adaptation strategy. Assume that the network topology is dense enough such that it is possible to find at least one neighbor unit around each unit along a variable route, and also that it is possible to find at least one substitute either at a global or local region. Though, these evaluations are based on an ideal operating environment, it still can reflect the essentials of our solution. The total latency of route switching at the local region from detecting a route failure to renewing the data message routing can be expressed as: Total latency (TL) = Latency (L_{REM}) for the time-out and the routing REM (the Route Error Message) + Latency (L_{RQM}) for the routing RQM + Latency (L_{RRM}) for the routing RRM. These three types of routing latency can be expressed individually by the following functions:

$$L_{REM} = T_{Time-out} + \sum_{N_{REM}=0}^{n} \left(T_{REM-P} \times N_{REM} \right) + \sum_{N_{REM}=0}^{n} \left(T_{REM-T} \times N_{REM} \right)$$

$$L_{RQM} = \sum_{N_{RQM}=0}^{n} \left(T_{RQM-P} \times N_{RQM} \right) + \sum_{N_{RQM}=0}^{n} \left(T_{RQM-T} \times N_{RQM} \right)$$

$$L_{RRM} = \sum_{N_{RRM}=0}^{n} \left(T_{RRM-P} \times N_{RRM} \right) + \sum_{N_{RRM}=0}^{n} \left(T_{RRM-T} \times N_{RRM} \right)$$

Here, $T_{Time-out}$ denotes the time-out latency for the route failure confirmation at the nth unit. N_{**M} denotes the number of units a message ** has to hop. $T_{(**M)-P}$ denotes the whole latency of a message processing needed at a unit. $T_{(**M)-T}$ denotes the latency of the message transmission over free space.

Also, assume that duplex communication is available between the units in this example network that allows sending a routing-error message backward directly to the source unit from a route unit that encounters a route failure. The RQM and the RRM will pass through the same units for route discovery processing. The message transmission over free space incurs little latency compared to that of the message processing at the units; hence it is ignored from our calculation. Further, assume that the latencies for all types of message processing (the REM, the RQM, the RRM and the LWM) are the same and are a constant for every unit in this network. Then, the total latency to renew the data message routing for units encountering route failure is:

$$TL_{Global} = T_{Time-Out} + \sum_{N_{REM}=0}^{n}\left(T_{(\cdot\cdot M)-P} \times N_{REM}\right) + \sum_{N_{RRM}=0}^{n}\left(T_{(\cdot\cdot M)-P} \times 2 \times N_{RRM}\right)$$

$$= T_{Time-Out} + T_{(\cdot\cdot M)-P} \times \left(\sum_{N_{RRM}=0}^{n} N_{REM} + 2 \times \sum_{N_{RRM-RQM}=0}^{n} N_{RRM-RQM}\right)$$

From the above function, it is clear that the values of N_{REM} and N_{RRM} will decide how much is the total latency. The N_{REM} variable is irrelevant for the case where the REM will not reach the source unit. However, the $T_{Time-out}$ setting at the source unit should allow the maximum duration for a message routing from the destination to the source unit. This duration is the same as that of the worst case of the REM passing through all hops: $T_{Time-out} = \Sigma(T_{REM-P} \times N_{REM})$, if the latency of the message transmission over free space has not been considered.

For the local adaptation algorithm without the advance warning (AW) mechanism, following the same assumptions, the total latency, TL_{Local}, to renew a data message routing at local region is $TL_{Local} = T_{\cdot\cdot M-P} \times \Sigma(N_{\cdot\cdot M})$. The $N_{\cdot\cdot M}$ variable decides the total latency needed to renew a data message routing, and it is a small number. Then, considering a minimum common hopping number that could satisfy the three possible migration models simultaneously, the common minimum HN = the maximum value of {2, 3, 2} = 3; The minimum hopping number that can satisfy these three possible migration models simultaneously is 3. In the local adaptation with the advance warning mechanism, the warning duration, $T_{warning\ duration}$, depends strongly on the operating environment. If this duration is set to be larger than the total latency needed to renew the data message routing, the total latency is zero, $TL_{Local/AW} = 0$, for the new local adaptation strategy with the advance warning. Thus, the minimum interval for the warning duration is:

$$TL_{Local/AW} = 0 = T_{warning\ duration} - T_{LWM-P} \times \Sigma(N_{LWM});$$
$$T_{warning\ duration} = T_{LWM-P} \times \Sigma(N_{LWM});$$
$$T_{warning\ duration} = T_{LWM-P} \times 3;$$

The minimum boundary value for setting the warning duration is equal to $T_{LWM-P} \times 3$. Figure 6 shows the results of these evaluations. From the results, it is clear that a routing algorithm without the local adaptation capability will spend considerable time on waiting for the route switching to be complete when a route failure occurs. By introducing our proposed algorithm, the efficiency for the route switching can get a great improvement if a local substitute is available. It is possible to avoid unnecessary searching for the local substitute if the local routing table at each unit has perfect maintenance. In other words, a route unit just needs to check its own local routing table when encountering route failure and know whether or not a neighbor unit exists around it. If there is no available neighbor unit, this unit can abandon the local adaptation attempt and send a warning message directly to the source unit. During the period of the warning message routing to the source unit, the initial route can continue the initial data message transmission. Thus, in the worst case, this proposed global adaptation algorithm could still reduce the latency for the route switching in a global sense.

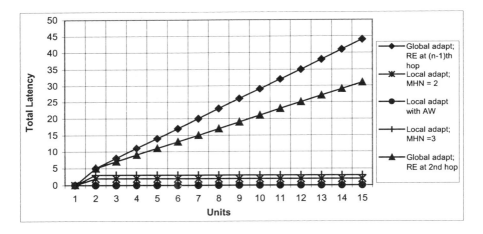

Figure 6. The total latency needed to renew the data message routing as the *(n-1)*th unit and the 2nd unit encounters route failure in a route with n units compared with that of local adaptation with advance warning capability.

As part of the future work, the essential characteristics of the mobile units of the Ad-hoc networks will be further studied. In a real world, the migration of mobile units that causes the unstable signal is common. Thus a routing algorithm for such network should have enough capabilities to adapt possible topology changes, no matter if the topology is stable or dynamic [25,26]. Through the investigation on these essential characteristics, a more robust and practical routing algorithm for the wireless mobile Ad-hoc network could be setup. Our research will consider the simulation and the determination of the unit throughput in the route switching using the proposed technique under a more realistic topological setting.

REFERENCES

[1] S. Chen and K. Nahrstedt, "Distributed Quality-of-Service Routing in Ad-hoc Networks," *IEEE Journal on Selected Areas in Comm.*, vol.17, no. 8, August 1999, pp. 1488-1505.

[2] T. Imielinski and H. F. Korth, *Introduction to Mobile Computing*, Kluwer Academic Publishers, 1996, pp. 1-43.

[3] B. Jabbari, G. Colombo, A. Nakajima, and J. Kulkarni, "Network Issues for Wireless Communications," *IEEE Comm. Magazine*, January 1995, pp.88-98.

[4] E. Pagani and G. P. Rossi, "Providing Reliable and Fault Tolerant Broadcast Delivery in Mobile Ad Hoc Networks," *Mobile Networks and Applications*, April 1999, pp. 175-192.

[5] M. Weiser, B. Welch, A. Demers, and S. Shenker, *Scheduling for Reduced CPU Energy*, Kluwer Academic Publishers, 1996, pp. 448-470.

[6] T. Imielinski, B. R. Badrinath, "Mobile Wireless Computing," *Communication of the ACM*, vol. 37, no.10, Oct. 1994, pp. 18-28.

[7] J. E. Padgett, C. G. Gunther, and T. Hattori, "Overview of Wireless Personal Communications," *IEEE Comm. Magazine* , Jan. 1995, pp. 28-41.

[8] V. K. Varma, P. W. Roder, M. Ulema, and D. J. Harasty, "Architecture for Interworking Data over PCS," *IEEE Comm. Magazine*, Sept. 1996, pp. 124-130.

[9] D. Bertsekas, R. Gallager, *Data Networks*. 2nd Ed., Prentice Hall. Ch. 5, 1992.

[10] J. M. McQuillan and D. C. Walden, "The ARPA Network Design Decisions," *Computer Networks 1*, North-Holland Pub. Co., January 1977, pp. 243-289.

[11] J. M. McQuillan, I. Richer, and E. C. Rosen, "The New Routing Algorithm for the ARPANET," *IEEE Transactions on Comm.*, vol. Com-28, no 5, May 1980, pp. 711-719.

[12] M. Schwartz and T. E. Stern, "Routing Techniques Used in Computer Communication networks," IEEE *Transactions on Comm.*, vol. Com-25, no. 4, pp. 539-552, April 1980.

[13] R.W. Floyd, Algorithm 97: Shortest Path, CACM 5(6), 345, June 1962.

[14] D. B. Johnson and D. A. Maltz, *Dynamic Source Routing in Ad Hoc Wireless Networks*, Kluwer Academic Publishers, 1996, pp.153-181.

[15] R. Dube, C. D. Rais, K. Y. Wang, and S. K. Tripathi, "Signal Stability-Based Adaptive Routing (SSA) for Ad Hoc Mobile Networks," *IEEE Personal Comm.*, February 1997, pp. 36-45.

[16] M. S. Corson, and A. Ephremides, "A Distributed Routing Algorithm for Mobile Wireless Networks," *ACM J. Wireless Networks*, January 1995, pp. 61-81.

[17] Y. L. Chang and C. C. Hsu, "Routing in Wireless/Mobile Ad Hoc Networks Via Dynamic Group Construction," *Mobile Networks and Applications*, May 2000, pp. 27-37.

[18] B. Das, R. Sivakumar, and V. Bharghavan, "Routing in ad hoc networks using a virtual backbone," *ACM SIGCOMM '97*, January 1997, pp. 1-20.

[19] P. Krishna, N. H. Vaidya, M. Chatterjee, and D. K. Pradhan, "A Cluster-Based Approach for Routing in Ad Hoc Networks," *Proc. of the Second USENIX*, 1995, pp. 1-10.

[20] C. E. Perkins and P. Bhagwat, "Highly Dynamic Destination-Sequenced Distance-Vector Routing (DSDV) for Mobile Computers," *Computer Comm. Review*, 24, 4 (ACM SIGCOMM 1994), 1996, pp. 234-244.

[21] G. E. Keiser, *Local Area Networks*, McGraw-Hill, Ch. 2, 1992.

[22] P. Lorrain and D. R. Corson, "Electromagnetic Fields and Waves," W. H. Freeman and CO., ch. 10 - 12, 1972.

[23] H. M. Hafezetal, "Fundamental Issues in Millimeter Wave Indoor Wireless Networks," Conference Record, *Wireless 1993 Conference*, Calgary, July 1993.

[24] J. G. Pickholtz, *Digital Communications,* McGraw-Hill, 1989.

[25] R. Katz, "Adaptation and Mobility in Wireless Information Systems," *IEEE Personal Comm.*, 1994, pp. 6-17.

[26] A. B. McDonald and T. F. Znati, "A Mobility-Based Framework for Adaptive Clustering in Wireless Ad Hoc Networks," *IEEE Journal on Selected Areas in Comm.*, vol.17, no. 8, August 1999, pp. 1466-1487.

Index